Hazards, Decontamination, and Replacement of PCB

A Comprehensive Guide

ENVIRONMENTAL SCIENCE RESEARCH

Series Editor:
Herbert S. Rosenkranz
Department of Environmental Health Sciences
Case Western Reserve University
School of Medicine
Cleveland, Ohio

Founding Editor:
Alexander Hollaender

A Continuation Order Plan is available for this series. A continuation order will bring delivery of each new volume immediately upon publication. Volumes are billed only upon actual shipment. For further information please contact the publisher.

Hazards, Decontamination, and Replacement of PCB

A Comprehensive Guide

Edited by
Jean-Pierre Crine

Hydro-Quebec Research Institute
Varennes, Canada

PLENUM PRESS • NEW YORK AND LONDON

Library of Congress Cataloging in Publication Data

Conference on PCBs and Replacement Fluids (1986: Montréal, Québec)
    Hazards, decontamination, and replacement of PCB: a comprehensive guide / edited by Jean-Pierre Crine.
        p.   cm. — (Environmental science research; v. 37)
    "Partially based on proceedings of the IEEE Montech '86 Conference on PCBs and Replacement Fluids, held September 29–October 1, 1986, in Montreal, Canada" — T.p. verso.
    Includes bibliographies and index.
    ISBN 0-306-43088-6
    1. Polychlorinated biphenyls — Toxicology — Congresses. 2. Polychlorinated biphenyls — Decontamination — Congresses. 3. Electric transformers — Health aspects — Congresses. I. Crine, Jean-Pierre. II. Title. III. Series.
    [DNLM: 1. Polychlorobiphenyl Compounds — congresses. W1 EN986F v. 37 / QV 633 C748h 1986]
RA1242.P7C66   1986
363.1'79 — dc19
DNLM/DLC   88-31630
for Library of Congress   CIP

Partially based on proceedings of the IEEE Montech '86 Conference on
PCBs and Replacement Fluids, held September 29–October 1, 1986,
in Montreal, Canada

© 1988 Plenum Press, New York
A Division of Plenum Publishing Corporation
233 Spring Street, New York, N.Y. 10013

Printed in the United States of America

Poly Chlorinated Biphenyls (PCBs) are dielectric liquids which have been widely used in various industries for more than 50 years because of their supposed nonflammability and their chemical inertness. Recent accidents all over the world have shown PCBs can burn and their combustion by-products (dioxines, furanes, etc.) are highly toxic. In fact, confusion has been created in the public mind between the dangers and hazards induced by PCBs themselves and those generated by their byproducts. Meanwhile, PCB pollution and toxicity is a major concern for regulating agencies, such as EPA in the United States and industry. Most Western countries now ban PCB production and strictly control their use. However, enormous amounts of PCB remain in use and their safe handling, destruction and replacement are heavy burdens for industrial users.

PCB pollution and its side effects are the subject of various studies with recent conferences devoted to these PCB studies. Thus a large body of specialized information now exists on the environmental, medical, bio-logical and safety aspects of PCB handling, use, cleaning and decontami-nation. However, no single comprehensive publication is yet available which deals with all the problems associated with PCBs. The major objective of the present book is to provide such a guide for PCB users.

One interest of this book is that it brings together the point of view of scientists from widely different backgrounds: biologists, physicians, environmentalists, toxicologists, chemists, electrical engineers, etc. The large spectrum of topics covered by the authors should enable anybody having to cope with a PCB incident to rapidly find practical solutions.

The main themes discussed in this book are: basic physicochemical proper-ties of PCBs, analytical techniques, human and environmental hazards, combustion byproducts, decontamination processes, destruction techniques and replacement fluids. Obviously, each section cannot claim to be complete but extensive reference lists are provided for readers needing more detailed information. We hope that this review of various PCBs problems and solutions will prove helpful to readers whatever their back-ground and field of interest.

It is a pleasure to thank all authors for their much appreciated contri-butions and for the diligence they have displayed in preparing their manuscripts.

Jean-Pierre Crine

Varennes, September 1987

CONTENTS

DESTRUCTION

COMBUSTION BY PRODUCTS AND
REPLACEMENT LIQUIDS

BASIC PROPERTIES
AND ANALYTICAL TECHNIQUES

POLYCHLORINATED BIPHENYLS (PCBs)

PHYSICAL AND CHEMICAL PROPERTY DATA

Douglas E. Metcalfe and George Zukovs
CANVIRO Consultants
A Division of CH2M HILL Engineering Ltd.
Waterloo and Mississauga, Ontario, Canada

Donald Mackay and Sally Paterson
Department of Chemical Engineering and Chemistry
University of Toronto
Toronto, Ontario, Canada

INTRODUCTION

CANVIRO Consultants, in association with others, have developed a polychlorinated biphenyls (PCBs) EnviroTIPS (Technical Information for Problem Spills) manual for Environment Canada (CANVIRO Consultants, 1987). The purpose of the PCB EnviroTIPS manual was to provide comprehensive information on PCBs for use by persons assessing the impacts of spills on the environment, developing spill countermeasures and formulating spill contingency plans. As such, PCB physical and chemical property data were an integral component of the manual, as these data are required for developing materials handling procedures, evaluating PCB movement and fate from spill sites with environmental transport models, and selecting appropriate spill countermeasures.

This paper summarizes the PCB physical and chemical property data assembled and developed for the PCB EnviroTIPS manual. A wide variety of PCB-containing fluids and solids exist, and as a result it was necessary to select specific types of PCB fluids and solids for physical-chemical characterization. Also, the PCB physical-chemical property data are fragmented, and data gaps exist, particularly for environmentally relevant properties. Estimates were made where necessary.

The paper describes the rationale for classifying and selecting types of PCB fluids and solids for characterization, and the methodology for estimating PCB properties where data gaps existed. The physical-chemical property data are then presented based on the classifications of PCB fluid and solid types.

PCB NOMENCLATURE

PCBs are a class of 209 discrete synthetic chemical compounds, called congeners, in which one to ten chlorine atoms are attached to biphenyl. The empirical formula for PCBs is thus $C_{12}H_{10-n}Cl_n$, where $n = 1$ to 10. The structural formula of the unsubstituted biphenyl molecule, with the numbering system of the carbon atoms in each ring, is as follows (Erickson, 1986):

The entire set of 209 PCBs forms a set of congeners. When PCBs are subdivided by the degree of chlorination (the parameter "n" in the above formula), the term isomer group is used. For example, the 24 different PCB compounds with three chlorine atoms are called the trichlorobiphenyl isomer group. PCBs of a given isomer group with different chlorine substitution positions are called isomers. Thus, 2,4,4'-trichlorobiphenyl and 2',3,4-trichlorobiphenyl are two of the 24 trichlorobiphenyl isomers. PCB nomenclature is summarized in Table 1.

CLASSES OF PCB-CONTAINING SOLIDS AND FLUIDS

The physical-chemical properties of PCBs vary on a congener-by-congener basis. Commercial PCB fluids are complex mixtures of these PCB congeners. Other chemical compounds were often added as diluents or additives to form commercial PCB-containing fluids. Additionally, various solids and fluids have been contaminated with PCB fluids and PCB-containing fluids.

Due to the wide variety of PCB-containing solids and fluids, it was necessary to make generalizations and approximations in selecting solids and fluids for physical-chemical property characterization.

The Monsanto Chemical Company was the sole producer of PCBs in North America, and the PCBs in Canada were imported almost exclusively by Monsanto. Additionally, the bulk of PCBs currently in use and in storage in Canada are or were associated with electrical applications, particularly capacitors and transformers (CANVIRO Consultants, 1987; Environment Canada, 1986). As such, physical-chemical property data were collected primarily for the commercial PCB fluids produced by Monsanto for electrical applications, and other associated PCB-containing and PCB-contaminated fluids.

Table 1.   PCB Nomenclature Categories

CATEGORY	NUMBER OF INDIVIDUAL COMPOUNDS
Congener	209
Isomer Group	10
Isomers/Group	1-46

Figure 1 compares the complete family of PCB and PCB-containing solids and fluids with those fluids selected for physical-chemical property characterization. Figure 1 is discussed below in terms of the categories or classifications of PCB and PCB-containing fluids selected for property characterization. The associated approximations that were employed in obtaining the physical-chemical data for the selected fluids are discussed in the next section.

PCB Isomer Groups

The unwieldy number of PCB congeners and the limited physical-chemical property data that are available for the individual congeners make the presentation of property data on a congener-by-congener basis impractical.

The compositions of commercial PCB fluids are often characterized in terms of the PCB isomer groups. In fact, congener property values are correlated with the number of chlorine atoms for most of the physical-chemical properties reported, with the nature and degree of correlation dependent on the property type.

Physical-chemical property data are, therefore, presented for the PCB isomer groups. Physical-chemical property data for biphenyl are also presented for completeness and because this compound is a constituent of some of the Monsanto commercial PCB fluids.

Selected Aroclors

Monsanto's commercial PCB fluids were manufactured under the "Aroclor" trade name. Three of these pure PCB fluids were used as capacitor Askarels, where "Askarel" is a generic name for synthetic, nonflammable, chlorinated aromatic hydrocarbon type electrical insulating fluids. As will be shown later, the various Aroclors can be defined in terms of their approximate compositions by isomer group.

Selected Transformer Askarels

PCB-containing transformer Askarels were manufactured as mixtures of Aroclors, chlorobenzenes and oxygen-scavenging additives.

PCB-Contaminated Mineral Oil

Mineral oils have in many circumstances become contaminated with PCBs, as a result of their usage in transformer applications.

METHODOLOGY FOR ESTIMATING PCB PHYSICAL-CHEMICAL PROPERTIES

As previously discussed, the physical-chemical property data for the individual PCB congeners are fragmented, with few properties defined for many congeners, and infrequent definition of the variations of these properties with temperature. As a result, the PCB isomer groups were selected as the fundamental or primary level for defining PCB physical-chemical properties. This reduced the number of fundamental PCB types to ten: monochlorobiphenyl, dichlorobiphenyl, trichlorobiphenyl, tetrachlorobiphenyl, pentachlorobiphenyl, hexachlorobiphenyl, heptachlorobiphenyl, octachlorobiphenyl, nonachlorobiphenyl and decaclorobiphenyl.

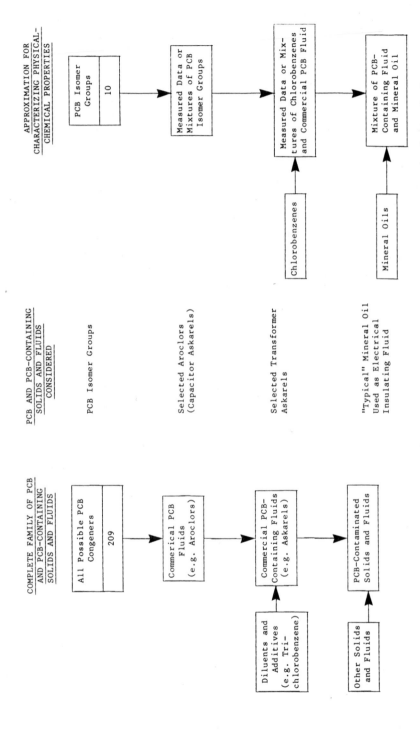

Figure 1. PCB and PCB-Containing Fluids and Solids Considered for Physical-Chemical Property Definition and Associated Approximations

6

Using interpolative methods and the available physical-chemical congener property data, average values for the properties were estimated for each of the ten PCB isomer groups at standard conditions. Appropriate temperature conversion techniques were used to modify these estimates for different temperatures.

Considerably more property data are available for the Aroclor PCB fluids than the PCB isomer groups and congeners, due to their commercial usage. However, there are gaps in the Aroclor physical-chemical property database, particularly for environmentally relevant properties. In these instances, appropriate mixing rules were used to estimate property values for the Aroclors, based on their average isomer group compositions and the PCB isomer group property values. Examples of the mixing rules that were employed are shown in Table 2.

As shown in Figure 1, the use of the mixing rules was extended to transformer Askarels in instances where property data were not available from the literature. For the transformer Askarels, the chlorobenzene diluents only were considered, as the additives constitute less than 0.25 percent by weight in all cases.

A computer program was written that incorporated the mixing rules and temperature conversion rules for all relevant physical-chemical properties. Physical-chemical properties computed with this program for Aroclors and Askarels with known property values were compared to those properties in order to validate the computer program, and appropriate modifications were made to the isomer group properties, temperature conversion factors and mixing rules as required. This reconciliation of published data and computed properties was made over a range of temperatures. Physical-chemical properties referenced as Mackay (1986) were calculated with the computer program.

Table 2. Mixing Rules Used to Calculate PCB Fluid Properties
(Reid et al, 1977)

PHYSICAL-CHEMICAL PROPERTY	MIXING RULE	EQUATION[1]	DEFINITION OF TERMS
Fluid Density	Additive Volume	$\rho_m = \sum\limits_{}^{n} V_i \rho_i$	ρ = mass density (kg/m^3) V = volume fraction
Vapour Pressure	Raoult's Law	$P_m = \sum\limits_{}^{n} x_i P_i$	P = vapour pressure (Pa) x = mole fraction
Absolute Viscosity	Logarithmic Additive Mole Fraction	$\log \mu_m = \sum\limits_{}^{n} x_i \log \mu_i$	μ = absolute viscosity $(Pa \cdot s)$
Solubility in Water	Raoult's Law	$s_m = \sum\limits_{}^{n} x_i s_i$	s = solubility in water (g/m^3)

1. Subscript m indicates a mixture property, whereas subscript i indicates a component property (n components in total).

TYPES AND USE OF THE PCB PHYSICAL AND CHEMICAL PROPERTY DATA

Data required by individuals in developing materials handling procedures, assessing PCB transport and fate from spill sites and selecting appropriate countermeasures are subdivided into three types: state properties, fluid properties and environmentally relevant properties. Key temperature-dependent parameters for modelling PCB transport from spill sites include fluid density and viscosity for movement through soil and in water, and vapour pressure for movement to the atmosphere. Figures are provided showing these three parameters as a function of temperature.

When estimating properties, it is best to identify the fluid in this paper which most closely resembles the spilled fluid. Some extrapolation or interpolation may be necessary. For dilute PCB - mineral oil mixtures, the fluid physical properties will approximate those of the mineral oil.

For ease of presentation, the tables and figures of PCB physical and chemical properties are compiled at the end of the paper.

It must be emphasized that the reported values of the physical-chemical properties are in many cases estimates and may contain error. In most cases this error is believed to be of little consequence when assessing the behaviour of PCB mixtures during handling. However, in critical exposure situations, appropriate care must be exercised when using these values.

PCB ISOMER GROUP PHYSICAL-CHEMICAL PROPERTIES

The compositions and number of isomers of the ten PCB isomer groups and biphenyl are shown in Table 3. Selected properties for biphenyl and the PCB isomer groups are presented in the following tables:

State Properties	- Table 4
Fluid Properties	- Table 5
Environmentally Relevant Properties	- Table 6

Isomer group properties as functions of temperature are presented in the following figures (Mackay, 1986):

Vapour Pressure	- Figure 2
Absolute Viscosity	- Figure 3

PHYSICAL-CHEMICAL PROPERTIES OF SELECTED AROCLORS (CAPACITOR ASKARELS)

The average compositions of selected Aroclors in terms of the PCB isomer groups are provided in Table 7. Aroclors 1242, 1254 and 1016 were used without modification as capacitor Askarels, and thus these Aroclors also have a capacitor Askarel designation (ASTM, 1984b):

Capacitor Askarel	Aroclor
Type A	1242
Type B	1254
Type D	1016

Table 3. Composition of PCB Isomer Groups

PCB ISOMER GROUPS	EMPIRICAL FORMULA	PERCENT CHLORINE	NO. OF ISOMERS
Biphenyl	$C_{12}H_{10}$	0	1
Monochlorobiphenyl	$C_{12}H_9Cl$	19	3
Dichlorobiphenyl	$C_{12}H_8Cl_2$	32	12
Trichlorobiphenyl	$C_{12}H_7Cl_3$	41	24
Tetrachlorobiphenyl	$C_{12}H_6Cl_4$	49	42
Pentachlorobiphenyl	$C_{12}H_5Cl_5$	54	46
Hexachlorobiphenyl	$C_{12}H_4Cl_6$	59	42
Heptachlorobiphenyl	$C_{12}H_3Cl_7$	63	24
Octachlorobiphenyl	$C_{12}H_2Cl_8$	66	12
Nonachlorobiphenyl	$C_{12}HCl_9$	69	3
Decachlorobiphenyl	$C_{12}Cl_{10}$	71	1

Erickson, 1986

Table 4. State Properties of PCB Isomer Groups

PCB ISOMER GROUP	PHYSICAL STATE (1) (4)	MELTING POINT (°C)(1)(3)	BOILING POINT (°C)(2)(3)	VAPOUR PRESSURE (Pa) at 25°C (2)(4)(5)
Biphenyl	S	71	256	4.9
Monochlorobiphenyl	S/L	25-77.9	285	1.1
Dichlorobiphenyl	S/L	24.4-149	312	0.24
Trichlorobiphenyl	S/L	28-87	337	0.054
Tetrachlorobiphenyl	S/L	47-180	360	0.012
Pentachlorobiphenyl	S/L	76.5-124	381	2.6×10^{-3}
Hexachlorobiphenyl	S/L	77-150	400	5.8×10^{-4}
Heptachlorobiphenyl	S/L	122.4-149	417	1.3×10^{-4}
Octachlorobiphenyl	S/L	159-162	432	2.8×10^{-5}
Nonachlorobiphenyl	S/L	182.8-206	445	6.3×10^{-6}
Decachlorobiphenyl	S	305.9	456	1.4×10^{-6}

Notes: (1) Varies with isomer, S is solid, L is liquid
(2) Average properties of all isomers in group
(3) Shiu and Mackay, 1986
(4) Mackay, 1986
(5) Mean value for liquid

Table 5. Fluid Properties of PCB Isomer Groups

PCB ISOMER GROUP	APPROXIMATE SPECIFIC GRAVITY	VISCOSITY	
		mPa.s at 25°C	Universal Saybolt s at 25°C
Biphenyl	1.0	17	78
Monochlorobiphenyl	1.1	20	80
Dichlorobiphenyl	1.3	28	100
Trichlorobiphenyl	1.4	56	190
Tetrachlorobiphenyl	1.5	200	610
Pentachlorobiphenyl	1.5	1.5×10^3	4.4×10^3
Hexachlorobiphenyl	1.6	2.9×10^4	8.2×10^4
Heptachlorobiphenyl	1.7	$>10^6$	$>10^6$
Octachlorobiphenyl	1.7	$>10^6$	$>10^6$
Nonachlorobiphenyl	1.8	$>10^6$	$>10^6$
Decachlorobiphenyl	1.8	$>10^6$	$>10^6$

Mackay, 1986

Table 6. Selected Environmentally Relevant Properties of PCB Isomer Groups

PCB ISOMER GROUP	MOLECULAR WEIGHT (g/mol) (1)	WATER SOLUBILITY at 25°C (g/m^3) (2)	LOG OCTANOL-WATER PARTITION COEFFICIENT (2)	APPROXIMATE BIOCONCENTRATION FACTOR IN FISH (2)	APPROXIMATE EVAPORATION RATE at 25°C (g/(m^2 h))(2)
Biphenyl	154.2	9.3	4.3	1000	0.92
Monochlorobiphenyl	188.7	4.0	4.7	2500	0.25
Dichlorobiphenyl	223.1	1.6	5.1	6300	0.065
Trichlorobiphenyl	257.5	0.65	5.5	1.6×10^4	0.017
Tetrachlorobiphenyl	292.0	0.26	5.9	4.0×10^4	4.2×10^{-3}
Pentachlorobiphenyl	326.4	0.099	6.3	1.0×10^5	1.0×10^{-3}
Hexachlorobiphenyl	360.9	0.038	6.7	2.5×10^5	2.5×10^{-4}
Heptachlorobiphenyl	395.3	0.014	7.1	6.3×10^5	6.2×10^{-5}
Octachlorobiphenyl	429.8	5.5×10^{-3}	7.5	1.6×10^6	1.5×10^{-5}
Nonachlorobiphenyl	464.2	2.0×10^{-3}	7.9	4.0×10^6	3.5×10^{-6}
Decachlorobiphenyl	498.7	7.6×10^{-4}	8.3	1.0×10^7	8.5×10^{-7}

Notes: (1) Mackay et al, 1983
(2) Mackay, 1986

Table 7. Average Composition of Selected Aroclors
(Percent by Weight)

PCB ISOMER GROUP	AROCLOR						
	1221	1232	1016	1242	1248	1254	1260
Biphenyl	11	6					
Monochlorobiphenyl	51	26	2	1			
Dichlorobiphenyl	32	29	19	16	2		
Trichlorobiphenyl	4	24	57	49	18	1	
Tetrachlorobiphenyl	2	15	22	25	40	21	
Pentachlorobiphenyl				8	36	48	12
Hexachlorobiphenyl				1	4	23	38
Heptachlorobiphenyl						6	41
Octachlorobiphenyl							8
Nonachlorobiphenyl							1
Decachlorobiphenyl							

NRC, 1979

Table 8. State Properties of Selected Aroclors

PROPERTY	AROCLOR					
	1016	1242	1248	1254	1260
Appearance [4]	Clear [3] Mobile Oil	Clear Mobile Oil	Clear Mobile Oil	Light Yellow Viscous Liquid	Light Yellow Soft Sticky Resin
Colour, APHA, Max [4]	40 [3]	100	100	100	150
Physical State [3][4] at 25°C, 1 atm	Liquid	Liquid	Liquid	Liquid	Solid
Pour Point (°C) [2]	-14 [3]	-19	-7	10	31
Vapour Pressure [1] at 25°C (Pa)	0.10	0.091	0.023	6.7×10^{-3}	6.4×10^{-4}

Notes: (1) Mackay, 1986
(2) Hutzinger et al, 1974
(3) ASTM, 1984b
(4) Monsanto

Table 9. Fluid Properties of Selected Aroclors

PROPERTY	AROCLOR				
	1016	1242	1248	1254	1260
Viscosity at 25°C (3) (mPa s) (Universal Saybolt s)	45 200	69 270	280 1.0×10^3	2.0×10^3 6.7×10^3	1.9×10^5 6.0×10^5
Approximate Specific Gravity at 25°C (3)	1.4	1.4	1.4	1.5	1.6
Fluid Density at 25°C (kg/m^3)(2)	1,370[1] (20°C)	1,381	1,445	1,539	1,621
Max. Moisture Content (ppm) (2)	35[4]	50	50	50	50

Notes: (1) Erickson, 1986
(2) Monsanto
(3) Mackay, 1986
(4) ASTM, 1984b

Table 10. Environmentally Relevant Properties of Selected Aroclors

PROPERTY	AROCLOR				
	1016	1242	1248	1254	1260
Molecular Weight (g/mol) [1]	256	261	297	327	375
Solubility in Water at 25°C (g/m^3)[1]	0.84	0.75	0.32	0.14	0.035
Log10 Octanol/Water Partition Coefficient (Log K_{ow})[1]	4.4–5.8	4.5–5.8	5.8–6.3	6.1–6.8	6.3–7.5
Bioconcentration Factor in Fish (BCF)[1]	1.3×10^3–3.2×10^4	1.6×10^3–3.2×10^4	3.2×10^4–1.0×10^5	6.3×10^4–3.2×10^5	1.0×10^5–1.6×10^6
Evaporation Rate at 25°C (g/m^2/h)[1]	0.031	0.029	8.3×10^{-3}	2.7×10^{-3}	2.9×10^{-4}
Flash Point [2] (COC) (°C)	>B.P. [3]	176–180	193–196	>B.P.	>B.P.
Fire Point [2] (COC) (°C)	>B.P.	>B.P.	>B.P.	>B.P.	>B.P.

Notes: (1) Mackay, 1986
(2) Monsanto
(3) ASTM, 1984b
COC Cleveland Open Cup
B.P. Boiling Point

Table 11. Composition of Transformer Askarels (ASTM, 1984a)
(percent by weight)

COMPONENT	ASKAREL						
	Type A	Type B	Type C	Type D	Type E	Type F	Type G
Aroclor 1260	60	45					
Aroclor 1254				70	100	45	60
Aroclor 1242			80				
Trichlorobenzene [1]	40						40
Tri-Tetra Blend [2]		55	20	30		55	
Phenoxypropene oxide [3]	0.18 to 0.22			0.18 to 0.22	0.18 to 0.22		
Diepoxide-Type Compound [3]		0.115 to 0.135	0.115 to 0.135			0.115 to 0.135	0.115 to 0.135

Notes: (1) Mixture of isomers of trichlorobenzene
(2) Mixture of isomers of trichlorobenzene and tetrachlorobenzene
(3) Oxygen scavenging compounds
  - Phenoxypropene oxide ~ Glycidyl phenyl ether
  - Diepoxide ~ 3,4 Epoxycycloheylmethyl -3,4
    epoxycyclohexane carboxyate

Table 13. Fluid Properties of Transformer Askarels

PROPERTY	ASKAREL						
	Type A	Type B	Type C	Type D	Type E	Type F	Type G
Absolute Viscosity[2] (mPa s) at 25°C	31			33			
Kinematic Viscosity[1][3] at 37.8°C (Saybolt Universal s)	54	43	53	59	87	41	46
Specific Gravity at 15.5°C[1][3]	1.564	1.566	1.419	1.523	1.387	1.520	1.510
Density at 25°C (kg/m³)[2]	1,563			1,524			
Max. Moisture Content[1] (ppm)	30	30	30	30	30	30	30

Notes: (1) ASTM, 1984a (Average Specification)
  (2) Mackay, 1986
  (3) Average Specification

Table 12. State Properties of Transformer Askarels

PROPERTY	ASKAREL						
	Type A	Type B	Type C	Type D	Type E	Type F	Type G
Appearance[2]	Clear Mobile Oil	Clear Mobile Oil	Clear Mobile Oil	Clear Mobile Oil	Clear Mobile Oil	Clear Mobile Oil	Clear Mobile Oil
Colour,[2] APHA, Max	150	150	150	150	150	150	150
Physical State[2] at 25°C, 1 atm	Liquid	Liquid	Liquid	Liquid	Liquid	Liquid	Liquid
Pour Point,[2] Max (°C)	-32	-44	-30	-30	-14	-42	-38
Vapour Pressure [1][3] at 25°C (Pa)	2.5×10^{-4}			1.5×10^{-3}			

Notes: (1) Vapour Pressure of PCB component of Askarel mixture
(2) ASTM, 1984a
(3) Mackay, 1986

19

Table 14. Environmentally Relevant Properties of Transformer Askarels

PROPERTY	ASKAREL						
	Type A	Type B	Type C	Type D	Type E	Type F	Type G
Molecular Weight (g/mol)[3]	258.6			263.4			
Water Solubility[1][3] at 25°C (g/m^3)	0.016			0.056			
Evaporation Rate[2][3] at 25°C ($g/m^2/h$)	7.8×10^{-5}			4.8×10^{-4}			
Fire Point[4] (COC) °C	>B.P.	>B.P.	>B.P.	>B.P.	>B.P.	>B.P.	>B.P.

Notes:
(1) Aqueous solubility of PCB component of Askarel mixture.
(2) Evaporation rate of PCB component of Askarel mixture.
(3) Mackay, 1986
(4) ASTM, 1984a
COC Cleveland Open Cup
B.P. Boiling Point

Table 15. Properties of Electrical Insulating Oils

CATEGORY	PROPERTY	VALUE/RANGE
State	Appearance	Clear and Bright[2]
	Colour, Max.	0.5[2]
	Physical State at 25°C and 1 atm	Liquid[2]
	Molecular Weight (Approximate) (g/mol)	230[3]
	Pour Point Max.(°C)	-40[1][2]
Fluid	Absolute Viscosity at 25°C (mPa s)	21[3]
	Specific Gravity 15°C/15°C, Max.	0.91[2]
	Density at 15°C (kg/m^3)	875[3]
Environmentally Relevant	Solubility in Water at 25°C (g/m^3)	2.5×10^{-4}[3]
	Minimum Flash Point (°C)	145[2]

Notes: (1) In certain areas of Canada, it is common practice to specify a higher or lower pour point, depending upon climatic conditions.

(2) ASTM, 1984c

(3) Mackay, 1986

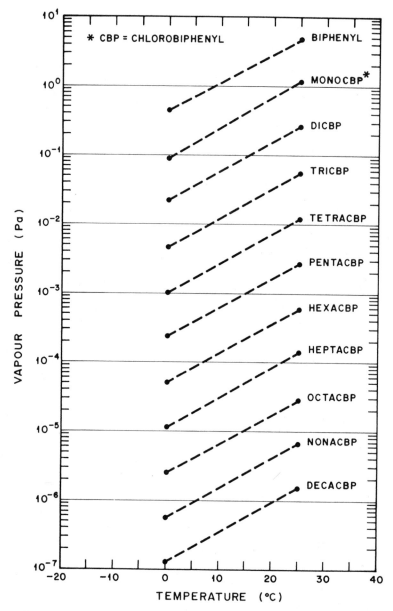

Figure 2.   PCB Isomer Group Vapour Pressure vs. Temperature

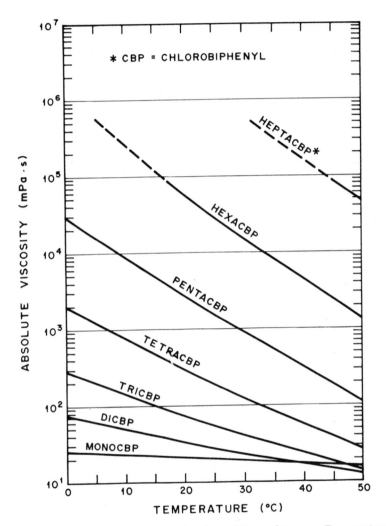

Figure 3. PCB Isomer Group Absolute Viscosity vs. Temperature

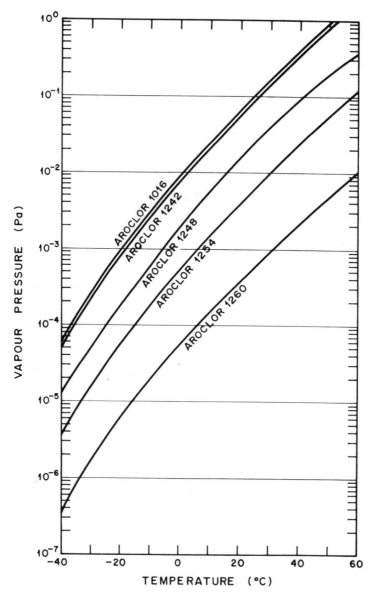

Figure 4.   Aroclor Vapour Pressure vs. Temperature

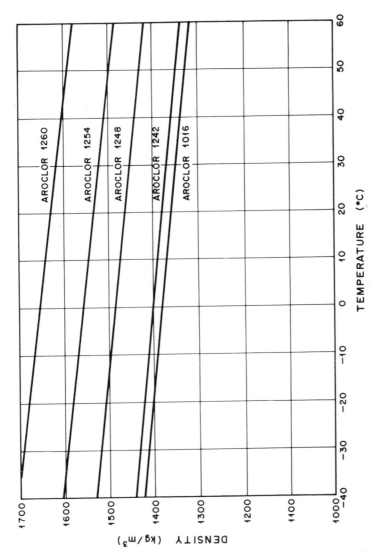

Figure 5. Aroclor Fluid Density vs. Temperature

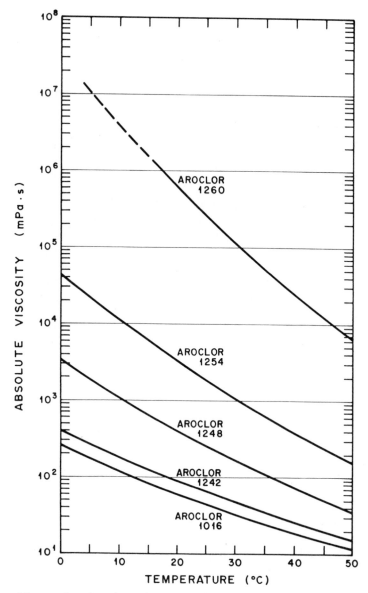

Figure 6. Aroclor Absolute Viscosity vs. Temperature

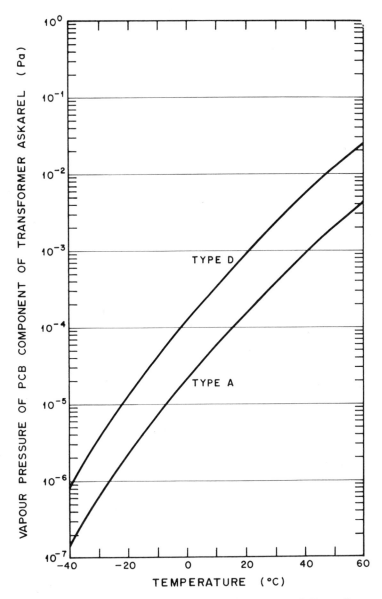

Figure 7. Vapour Pressure of PCB Component of Transformer
Askarel vs. Temperature

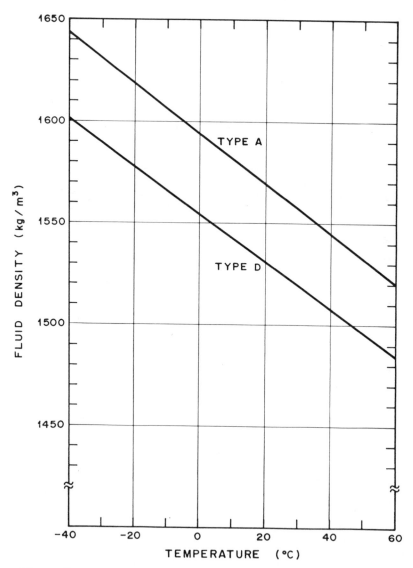

Figure 8. Transformer Askarel Fluid Density vs. Temperature

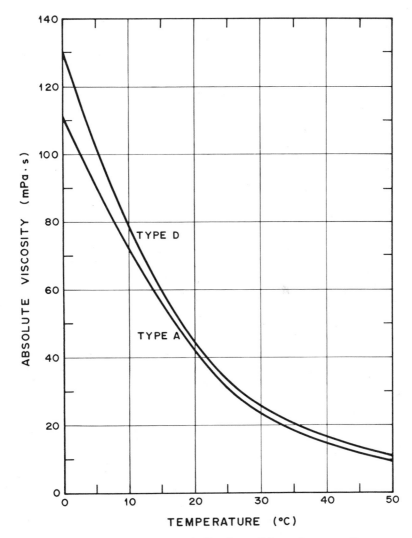

Figure 9.   Transformer Askarel Absolute Viscosity vs. Temperature

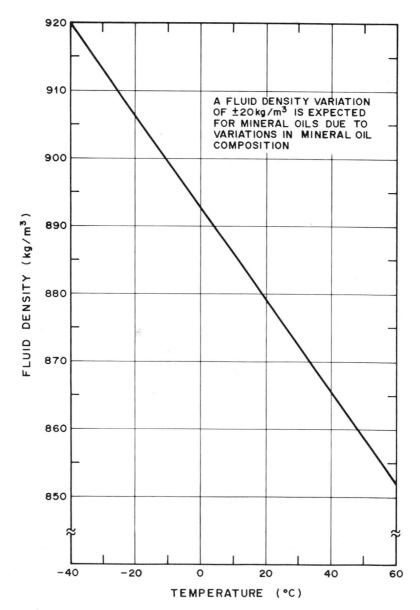

A FLUID DENSITY VARIATION OF ±20kg/m³ IS EXPECTED FOR MINERAL OILS DUE TO VARIATIONS IN MINERAL OIL COMPOSITION

Figure 10. Mineral Oil Fluid Density vs. Temperature

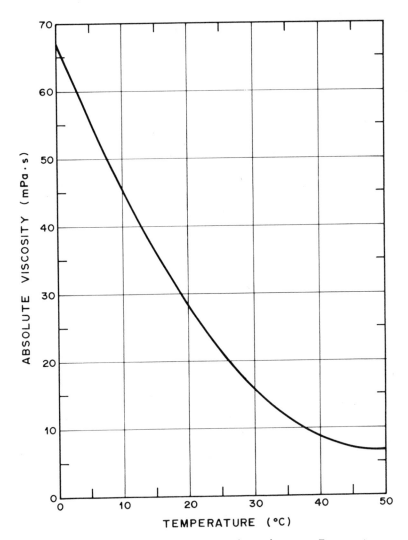

Figure 11. Mineral Oil Absolute Viscosity vs. Temperature

Selected Aroclor physical-chemical properties are presented in the following tables:

State Properties	- Table 8
Fluid Properties	- Table 9
Environmentally Relevant Properties	- Table 10

Aroclor properties as a function of temperature are presented in the following figures (Mackay, 1986):

Vapour Pressure	- Figure 4
Fluid Density	- Figure 5
Absolute Viscosity	- Figure 6

PHYSICAL-CHEMICAL PROPERTIES OF SELECTED TRANSFORMER ASKARELS

The average compositions of selected transformer Askarels are provided in Table 11. Trade names for these transformer Askarels are as follows:

Trade Name	Transformer Askarel
Pyranol (General Electric)	Type A
Inerteen (Westinghouse Electric)	Type D

Selected transformer Askarel physical-chemical properties are presented in the following tables:

State Properties	- Table 12
Fluid Properties	- Table 13
Environmentally Relevant Properties	- Table 14

Transformer Askarel properties as a function of temperature are presented in the following figures (Mackay, 1986):

Vapour Pressure (PCB component of Askarel mixture)	- Figure 7
Fluid Density	- Figure 8
Absolute Viscosity	- Figure 9

PHYSICAL-CHEMICAL PROPERTIES OF MINERAL OILS

Selected mineral oil properties are presented in Table 15. Mineral oil fluid density and absolute viscosity as functions of temperature are presented in Figures 10 and 11, respectively (Mackay, 1986).

The PCB vapour pressure of a PCB-contaminated mineral oil is a function of PCB type, PCB mole fraction and temperature. Thus, numerous calculations and figures would be required to show PCB vapour pressures for contaminated mineral oils containing Aroclors and/or transformer Askarels. Where PCB vapour pressures for PCB-contaminated mineral oils are required for assessing PCB atmospheric concentrations, the reader can obtain an approximate value by multiplying the PCB weight or volume fraction in the contaminated fluid by the corresponding pure-form PCB vapour pressure for the given Aroclor/transformer Askarel contaminant, as provided in the previous sections.

ACKNOWLEDGMENTS

This research was funded by the Technical Services Branch, Environmental Protection Service, Environment Canada. Mr. Mervin F. Fingas was the Environment Canada Project Officer; and his assistance with this work is gratefully acknowledged.

REFERENCES

American Society for Testing and Materials, 1984a, Standard Specification for Chlorinated Aromatic Hydrocarbons (Askarels) for Transformers, in: "1984 Annual Book of ASTM Standards," Vol. 10.03, No. D2283-82, pp. 415-417, Philadelphia, Pennsylvania.

American Society for Testing and Materials, 1984b, Standard Specification for Chlorinated Aromatic Hydrocarbons (Askarels) for Capacitors, in: "1984 Annual Book of ASTM Standards," Vol. 10.03, No. D2233-80, pp. 412-414, Philadelphia, Pennsylvania.

American Society for Testing and Materials, 1984c, Standard Specification for Mineral Insulating Oil Used in Electrical Apparatus, in: "1984 Annual Book of ASTM Standards," Vol. 10.03, No. D3487-82a, pp. 519-523, Philadelphia, Pennsylvania.

ASTM. See American Society for Testing and Materials.

CANVIRO Consultants, 1987, "PCB EnviroTIPS Manual," in preparation, Technical Services Branch, Environmental Protection Service, Environment Canada, Ottawa.

Environment Canada, 1986, "National Inventory of Concentrated PCB (Askarel) Fluids (1985 Summary Update)," EPS5/HA/4, Commercial Chemicals Branch, Environmental Protection Service, Ottawa.

Erickson, M.D., 1986, "Analytical Chemistry of PCBs," Ann Arbor Science, Butterworth Publishers, Stoneham, Massachusetts.

Hutzinger, O., S. Safe, and V. Zitko, 1974, "The Chemistry of PCBs," CRC Press, Cleveland, Ohio.

Mackay, D., 1986, Personal Communication.

Mackay, D., S. Paterson, S.S. Eisenreich, and M.S. Simmons (Eds.), 1983, "Physical Behavior of PCBs in Great Lakes," Ann Arbor Science, Ann Arbor, Michigan.

Monsanto, "Aroclor Plasticizers," Technical Bulletin O/PL-306, Organic Chemicals Division, St. Louis, Missouri (No Date).

National Research Council, 1979, "Polychlorinated Biphenyls," National Academy of Sciences, Washington, D.C.

NRC. See National Research Council.

Reid, R.C., J.M. Prausnitz, and T.K. Sherwood, 1977, "The Properties of Gases and Liquids," 3rd Edition, McGraw-Hill, New York.

Shui, W.Y. and D. Mackay, 1986, A Critical Review of Aqueous Solubilities, Vapour Pressures, Henry's Law Constants and Octanol/Water Partition Coefficients for PCBs, J. Phys. Chem. Ref. Data, 15:911.

GAS CHROMATOGRAPHIC ANALYSIS OF PCBs

Robert Grob, J. Mathieu and H. Ricau

Laboratoire de chimie analytique appliquée
École nationale supérieure de chimie de Toulouse
31077 - Toulouse, France

INTRODUCTION

Polychlorinated biphenyls (PCB) have been analysed for many years in various mediums such as water, sediments, soils and biological mediums. More recently, oil has been added to this list with the implementation of regulations on transformer and reclaimed oils. However, the methods used to analyse PCBs in these oils is not the same as for other mediums since the similarity between the properties of PCBs and the oils could interfere with analysis results.

There are several analytical methods available for determining and testing PCBs[1-10]: IR spectroscopy, liquid chromatography (HPLC), mass spectrometry (MS) and gas chromatography (GC), among others. There exists also some simpler identification tests based on colorimetry that can detect organic chlorine.[11]However, these tests, though useful for a gross detection of PCBs (especially in mineral oil), are generally not able to determine whether the detected chlorine comes from PCB or from any other chlorinated substance.

Gas chromatography is clearly recognized now as the best method for analysing PCBs[1-3,6-9]. Gas chromatography has earned its reputation for several reasons including:

- the separation of the different compounds and possible recognition of the type of PCB based on the chromatogram;

- the detectors available are highly sensitive;

- there is little interference especially if the sample has been cleaned up; quantitative determination thus gives accurate results.

_____

This article first describes the procedures and the apparatus used and then specifies the particular methods depending on the variety of combinations possible, such as mixed families of PCBs, mineral oils and other insulating liquids and analysis in water, soil or other materials.

CHOICE OF CHROMATOGRAPHIC PARAMETERS

The chromatographic column

Current ASTM[4] and IEC[5] standards on GC analysis of PCBs refer to packed (see Fig. 1) and semi-capillary columns. However, in light of recent progress accomplished with capillary GC columns, the authors consider that these standards should be revised. Capillary columns present the following advantages (compare Figs. 1 and 2):

- the large number of plateaus, which reflects the powerful separating capacity, eliminates much interference especially if the sample was cleaned up;

- the type of PCB can be more easily recognized for the same reason since the chromatographic trace is much more detailed;

- for a detector of equivalent quality, the detection limit is lowered since the signal to noise ratio is improved.

A capillary column can also be chosen based on the application[1,6,7]. For instance, the goal may be a very detailed chromatogram to bring out certain compounds independently as in dioxine research, or a detailed chromatogram to determine a mixture of a variety of PCB families or a less detailed one for the assessment of a single PCB family. In fact, a high-performance column is not necessarily useful in all circumstance, which means that other criteria can be taken into consideration such as the time and cost of the analysis and the cost of the column.

Depending on the circumstances therefore, capillary columns with the following specifications can be used:

- length: 15 to 60 m;

- inside diameter: 0.2, 0.3 mm up to 0.5 mm; 0.1-mm diameter columns can now be considered;

- thickness of stationary phase: depending on the performance sought, values of 0.1 to 0.3 m will provide fully adequate results;

- nature of the stationary phase: same as for packed columns; low polar silicone-based phases are suitable (e.g., OV1, OV101, SE52, SE54) and are grafted depending on analysis temperature conditions. More polar phases could be used to obtain a more detailed analysis, however such phases usually have stricter temperature limits. It is interesting to note that with nonpolar phases, compounds are eluted in an order corresponding to the their respective boiling points.

Depending on the type of analysis and the performance aimed at, the best compromises should be sought with respect to column length and inside diameter and to phase thickness and nature. Although a highly-capillary column (e.g., 0.1 mm) may provide very detailed results, it could get overloaded quickly and result in an insensitive analysis. In such cases, a column cleanup should be considered since it is often overloaded by substances other than PCBs.

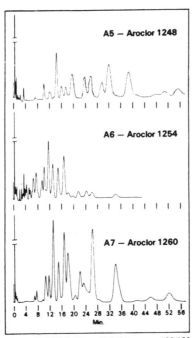

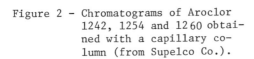

1.5% SP-2250/1.95% SP-2401 on 100/120 Supelcoport, 2.0m x 1.4" OD x 4mm ID TightSpec™ glass, Col. Temp.: 160°C (A1-A5) or 200°C (A6 & A7), Flow Rate: 60ml/min., N₂, Det.: ECD, Range: 10⁻¹², Attn.: 1 x 64, Sample: 5ng each Aroclor in 5μl isooctane.

Figure 1 - Chromatograms of Aroclor 1248, 1254 and 1260 obtained with a packed column (from Supelco Co.).

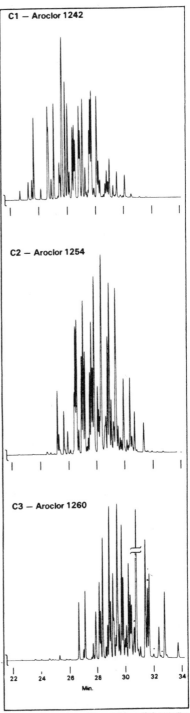

Figure 2 - Chromatograms of Aroclor 1242, 1254 and 1260 obtained with a capillary column (from Supelco Co.).

SPB-5 capillary column, 30m x 0.32mm ID fused silica, Film Thickness: 0.25μm, Col. Temp.: 30°C for 4 min., then to 300°C at 10°C/min. and hold 10 min., Linear Velocity: 25cm/sec., He, set at 150°C, Det.: ECD, Range: 10⁻¹¹, Attn.: 1 x 128, Sample: 1μl isooctane containing 1ng Aroclor standard; splitless injection.

Note: A precolumn (i.e., capillary column with no wall coating), which can protect the capillary column from certain impurities, often provides interesting results and does not hinder the separating capacity of the column.

The chromatographic detector

Three types of chromatographic detectors can be considered for analysing PCBs.

- the Hall-conductivity detector can be very specific with halogenated substances such as PCBs but is not always sensitive enough. Furthermore, it must be used very carefully and is not widely manufactured;

- the mass spectrometer directly coupled with a gas chromatograph (GC-MS) is a universal and sensitive detector which, when computerized, makes the identification of compounds possible even in the presence of other bothersome compounds (search for characteristic ions). Owing to its high cost, which can be a deterrent, this detector is used mainly for fine analyses rather than for routine work;

- the electron capture detector (ECD) is widely used for determining PCBs. This detector, which is specific for electronegative compounds, is very sensitive, responding to less than 0.1 pg of lindane, and is also relatively easy to use. It is ideal for routine PCB analysis. It uses a make-up gas such as pure nitrogen free of O_2 and H_2O or an argon-methane mixture. The latter mixture is recommended for extra sensitivity, such as for biological medium analysis, in which case the detector's linear range is increased. This detector can be maintained relatively easily. If it is used extensively, it may be necessary to dismantle and clean the collecting electrode. However, a simple and effective method consists in the hot chemical reduction of compounds adsorbed in the sensitive part of the detector using a flow of hydrogen; the make-up gas is replaced by hydrogen. The sensitivity of such detectors can be inconsistent especially after cleaning. Consequently, standard mixtures should often be injected in order to quantify the response factor as well as the linear range of the device (see Fig. 3).

The chromatographic injector

PCBs do not volatize easily. For example, at atmospheric pressure, the distillation range of PCBs chlorinated at 42% (Aroclor 1242) is from 325° to 336°C. For Aroclor 1254, it is from 365 to 390°C and for Aroclor 1260, from 385 to 420°C. This is another factor favoring the use of capillary columns which are better suited than packed columns for eluting non volatile compounds.

For a splitless injection, the temperature will be at least 250°C. However, the cold on-column injector seems to be best suited for the following reasons:

- the sample is completely injected and pollution of the injection port is avoided;

- injections can be easily repeated; quantitatively, the external standard method could be used.

Component (Quantity)

1 2-Chlorobiphenyl (100ng)
2 3,3'-Dichlorobiphenyl (100ng)
3 2,4,5-Trichlorobiphenyl (10ng)
4 2,2',4,4'-Tetrachlorobiphenyl (10ng)
5 2,3',4,5',6-Pentachlorobiphenyl (10ng)
6 2,2',3,3',6,6'-Hexachlorobiphenyl (10ng)
7 2,2',3,4,5,5',6-Heptachlorobiphenyl (5ng)
8 2,2',3,3',4,4',5,5'-Octachlorpbiphenyl (5ng)
9 2,2',3,3',4,4',5,5',6-Nonachlorobiphenyl (5ng)
10 2,2',3,3',4,4',5,5',6,6'-Decachlorobiphenyl (5ng)

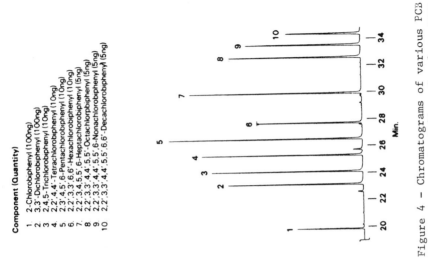

Figure 4 – Chromatograms of various PCB standard (from Supelco, same conditions as in Fig. 2).

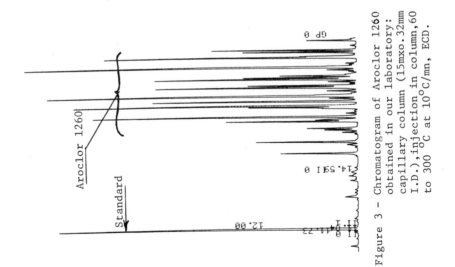

Figure 3 – Chromatogram of Aroclor 1260 obtained in our laboratory: capillary column (15mx0.32mm I.D.), injection in column,60 to 300 °C at 10°C/mn, ECD.

39

The injection temperature with this type of injector, which is in fact the solvent temperature when it leaves the chromatographic furnace since the injector is cold, is a function of the solvent used (e.g., 10°C below the solvent boiling point, followed by a temperature increase to obtain initial vaporization which is known as the solvent effect).

Carrier gas temperatures

Helium is a good carrier gas since it combines high separation performance and safety. The flow rate is a function of the inside diameter of the column used; for a 0.3-mm dia column, a flow rate of the order of 2 ml/min would be adequate.

The furnace heating schedule depends on the type of injector used, however for proper elution of PCBs, it is necessary to reach 250°C or even 300°C at a rate of increase of 5 to 15°C/min.

For maximum sensitivity, the temperature of the ECD must be in the area of 300°C inasmuch as the chromatographic column can withstand it. If the gas chromatograph is coupled with a mass spectrometer (GC-MS), the interface must be maintained at 250°C or higher. For analysis of PCTs (polychlorinated terphenyls), the above reference temperatures must be even higher.

Note: If a high-quality chromatograph and a very high-quality furnace are being used and if the injection protocol is so strict that it resembles that of an automatic injector, PCB retention times can be easily reproduced which is helpful for the recognition of chromatogram traces. A relative fluctuation of retention times of the order of 1/1000 can be obtained.

Choice of solvent

No solvent can be described as being incompatible with both the capillary column and the ECD. It is possible to use hydrocarbons such as petroleum ether, hexane, heptane, isooctane and aromatic hydrocarbons, cholorinated solvents such as dichloromethane, tetrachloroethylene or choroform, acetonitrile, dimethylformamide, ether, ethyl acetate, acetone, alcohols and so forth. The solvent should have the following properties:

- relatively low boiling point;

- high level of purity;

- high PCB solubility.

These solvents are used to extract the PCBs by liquid-liquid or liquid-solid extraction from their sample matrix (i.e., water, soil) or simply for dilution purposes prior to injection.

ANALYSIS OF A SINGLE TYPE OF PCB

Choice of standard

Many standard commercial PCBs (Fig. 4) are now available generally in the form of kits of various mixtures with the Aroclor nomenclature 1241, 1248, 1254 and 1260, which are the main families found in France depending on the applications (i.e., transformers, capacitors, distribution cells). Aroclor 1260, for example, is a mixture of PCBs representing those found in transformer pyralene -- 60% PCBs and 40% trichlorobenzenes by mass (Fig. 2-C3 and Fig. 4).

Analysis of an Aroclor 1260 solution and of a Pyralene T solution, both of which contain 60% of PCBs by mass and thus have the same level of concentration, yielded equivalent results with an ECD. The term "equivalent" is explained further on in this article.

When the quantity of PCBs is to be determined, it is unnecessary to identify each chromatographic peak since the ECD has very different response coefficients for PCBs that do not have the same number of chlorine atoms. However, if the number of chlorine atoms reaches or exceeds 3, this difference is attenuated (Fig. 3).

The quantity of a given type of PCB, such as Aroclor 1260, can therefore be expressed in ppm. A wide variety of tests conducted to determine quantities of PCBs have shown that even though the chromatogram is not identical to that of the standard Aroclor used, the accuracy of the analysis was not jeopardized. It should be noted that the difference in chromatograms refers to the different relative heights and not to the absence of peaks.

This observation about accuracy of analysis holds for the Aroclor 1254 and 1260 families of PCBs, but for those of the Aroclor 1242 family, it is not so obvious since, in all, more than 50% of the PCBs present would be mono-, di- and trichlorinated which have very different responses on the ECD. In contentious cases, it would be better to have standard solutions with one kind of PCB present, such as those marketed by Monsanto Co. and specialized chromatography companies like Supelco[8].

Calculation

It should be remembered at this point that we are interested here in determining the quantities of PCBs from one family and not from a mixture of families. The latter case is dealt with later in this article.

Two distinct possibilities should be mentioned.

1. The chromatogram representing the PCBs is comparable to that of a standard Aroclor. This is the easiest case.

   A PCB solution is compared directly with a standard Aroclor type solution. The sample is cleaned up and diluted as required. An internal standard compound, which also could be called an injection monitor*, is added to both equally-concentrated solutions.

   The chromatographic integrator is programmed to calculate the sum of the chromatographic areas of the PCBs in both solutions. The same

---

* Even for an on-column injection, the internal standard should be added for accuracy purposes. A statistical study on standard Aroclor showed that a 6% fluctuation without the internal standard compound could be lowered to 1.5% with it.

A wide choice of compounds are available, since the only selection criteria are that the ECD responds well and the retention time is different from that of the PCBs under study. Lindane (hexachlorocyclohexane) was used in this study, but many other compounds are suitable.

solvent is used for both solutions. PCB content can then be expressed in the following manner:

$$\text{ppm PCB} \atop \text{(unknown solution)} = \text{ppm Aroclor} \atop \text{(standard solution)} x A_I/A_E x A'_E/A'_I \qquad (1)$$

where:

A_I, A_E: the cumulative areas of the unknown PCB peaks and the Aroclor PCB peaks respectively.

A'_I, A'_E: the areas of the internal standard compound in the unknown solution and the standard solution respectively.

Note that the peak height can be substituted to the area but this requires some care in the analysis of the chromatogram. In order to obtain the content in the initial sample, the preparations made to obtain the solution that was tested before must be taken into account. These include the dilution or concentration factor, the original mass or volume prior to the liquid-liquid or liquid-solid extraction operations and so forth.

2 . If the chromatogram for the PCBs is different from that of a standard Aroclor compound, two explanations are possible.

- Additional peaks due to impurities, either with or without the PCB-retention times, in which case the PCBs are partially hidden; these impurities could possibly be eliminated by a better cleanup.

It sould be mentioned that if the part of the chromatogram of the PCBs undisturbed by the impurities is correct, the accuracy of the results will still be acceptable, however this part of the chromatogram must represent at least half of the overall printout.

The same equation is used with A'_I and A'_E representing here the cumulative surfaces of the unknown PCB peaks and the standard Aroclor compound peaks respectively, but which are limited in this case to the undisturbed part of the chromatogram. This part must be defined on the chromatogram of the unknown solution.

Should the chromatogram remain seriously disturbed because of impurities not removed during the cleanup, the analysis results will be, at best, an approximation. Recourse to a mass spectrometer coupled with a high-performance computer system, which might remove the interference, could then be considered. Use of a more efficient column in an aim to obtain a better discrimination between PCBs and impurities could also be considered.

- No additional peaks due to impurities but an incorrect chromatogram in which the similarity between the chromatograms of the unknown and standard solutions is unconvincing (peaks missing, etc.). We already pointed out that if the chromatogram of the solution under study does not have exactly the same form as that of the standard solution, the accuracy of the test, at least for the "heavy" PCBs of the 1254 or 1260 type, is not necessarily compromised.

When the chromatogram is too disturbed, individual standard compounds can be used. Mixtures of PCBs can be developed in which eachPCB represents a class (e.g., monochlorinated, dichlorinated,trichlorinated, etc.) and it can be assumed that isomers of aparticular class have similar responses. Under such conditions, equation (1) can be applied again with the chromatographic integration being broken down into sections that best represent the various classes. Consequently, A'_I and A'_E will represent respectively the surfaces of the unknown PCB peaks and the standard compound peak in a given section i, representing a class of PCBs (class = number of chlorinated substitutes). Results will be expressed in ppm of PCBs with a breakdown according to class and not acccording to a given Aroclor family.

Note: To obtain very accurate results, certain analyses can become very complex considering the wide variety of techniques available. Consequently, it is useful to define an acceptable level of accuracy that corresponds to the conclusion aimed at. The laboratory conducting the analyses would be well advised, however, to acquire data from the detector on a disc (IBM or equivalent) rather than on a simple integrator. This will enable it to process the chromatogram at will, file it, compare it, and so forth, ans also to obtain more accurate final results in less time.

When should mass spectrometry be used?

The ability and advantages of a mass spectrometer as a detector coupled with a gas chromatograph were presented above. If such a system is not accompanied by high-performance software and a library program, it is not particularly useful for analysing PCBs. Such a detector is much too cumbersome for routine analyses and since it is less sensitive in detecting PCBs than an ECD, it is illogical to use it.

On the other hand, it can be successfully used to remove the vagueness of more complex cases by the acquisition of the signal of certain characteristic ions, such as when PCB and standard compound chromatograms are different[9,10].

ANALYSIS IN THE PRESENCE OF A MIXTURE OF PCB FAMILIES

Some analyses reveal cases of mixtures of PCB families with varying degrees of complexity.

-   When two distinct families are present (i.e., Aroclor 1242 and 1260 types), both will have their own chromatogram trace undisturbed by the other. Only the intermediate section can be considered as the sum of the two PCB families.

    If the undisturbed 1242 and 1260 traces are similar to those of their respective standard compounds, the quantities can be determined as described for the single-family PCBs. The intermediate section must be compatible with the results.

-   When the two PCB families are not easily distinguishable (i.e., Aroclor 1254 and 1260 types), the method described above should be ruled out for precise measurements since the overlapping of the chromatogram traces is too great. Under these circumstances, results should be expressed in the form of ppmm (PCBs) based on the principle by which the integration is performed by section and the response coefficient established by class of PCB.

Note: In the event the sample is a mixture of families and also contains impurities despite the cleanup, the analysis can become very complex at which time mass spectrometry may be the answer.

ANALYSIS OF PCBs IN INSULATING OILS

The analysis of PCBs in mineral oils may appear initially to be problematic since the presence of the oil could seriously interfere unless the sample undergoes an appropriate cleanup. Considering the respective distillation ranges of oil and PCBs, chromatographic separation of the two is virtually impossible. The response of the ECD can be quenched by the presence of the mineral oil; the author has verified this. Some authors therefore recommend that the sample be prepared. For example, a hexane/acetonitril mixture, high in acetonitrile, can be used to separate the oil and the PCBs, with the latter going into the hexane, which is then purified on a florisil, silica or alumina cartridge. The purification operation is described below. These operations have the added advantage of eliminating most of the products of the oxidization of oil, which can hinder detection due to additional peaks or peaks interfering with those of the PCBs being analysed.

Dilution of the sample

If the sample has been purified to eliminate the oil, it can be diluted to any degree since the sensitivity is excellent – much less than 1 ppm of PCB in oil should be detected.

If no oil is eliminated, the sample must be considerably diluted. Even if the oil PCB content is minimal, the sample should be diluted with a solvent such as heptane at a factor of 100 to 200, or else the quenching factor will modify the detector response. A dilution of this magnitude performed on samples with a low PCB content (e.g., < 20 ppm) results in a high sensitivity on the ECD.

Impact of other products

This problem was mentioned previously, however, here the most bothersome parasite product is the oil and certain products of its oxidization. The elimination of oil was also dealt with above.

Mineral oil has almost no effect on the ECD, but since the PCBs arrive in the detector at the same time as the oil molecules, PCB detection is altered. Experience shows, in fact, that if the sample is diluted at a factor of at least 100 or 200 or even much more if PCB content is very high, the impact of the oil is minimized. Samples with a PCB content in oil ranging from 15 to several thousand ppm and more have been successfully tested without any prior preparation other than dilution. Nevertheless, in the interests of accuracy, the oil concentration in the standard Aroclor compounds approximated that of the unknown solutions. Should the PCB content be below 20 ppm, it would be not only useful but even essential to eliminate the oil, especially if it is oxidized.

Foregoing the cleanup stage may save time, but to do so gives rise to other problems:

- sensitivity limits suffer;

- column life is shorter (a precolumn that is regularly changed can compensate for this problem);

- the ECD must be cleaned more frequently;

Sensitivity and accuracy limits

If the sample has been cleaned up and is free of oil, the sensitivity is excellent as described above. On the other hand, for an untreated sample, the analysis is sensitive to a minimum PCB content of 1 to 2 ppm.

Accuracy depends on several factors:

- the quality of the PCB chromatogram (impurities, mixtures, ambiguous chromatogram, etc.);

- analysis statistics (number of injections);

- precision of standard compounds used.

The analyst should ascertain the degree of accuracy in a number of particular cases.

- A high degree of accuracy can easily be obtained with high PCB-content oils (i.e., less than 2% error in favorable circumstances). It should be recognized that a very high degree of accuracy in such cases is unnecessary since the critical point of 100 ppm is greatly exceeded;

- accuracy must be great when PCB-content approaches critical limits such as 50 ppm (reclaimed oils) and 100 ppm (transformer liquids). A statistical study performed showed that a degree of accuracy within 5% can be easily guaranteed at 100-ppm content levels with an average of three chromatographic injections. On the other hand, at 50 ppm and at 15 ppm, averages of five or six and 12 injections respectively would be required. The samples were not cleaned up in this study.

Purification of low PCB-content samples increases the accuracy of the integration - the base line and chromatogram are of higher quality - and, in general makes the analysis more reproducible. We, however, do not believe the analysis is necessarily more accurate since the PCB-loss factor resulting from the cleanup comes into play even though the standard samples undergo the same treatment. This question warrants a statistical study.

Other insulating oils

The remarks about mineral oils apply also for silicone oils although the quenching effect is almost inexistent. Nevertheless, the procedures followed in the preparation of mineral oils aimed at removing the oils should also be used to prepare these samples.

ANALYSIS OF PCBs IN DIFFERENT MEDIUMS

The literature has dealt with and described extensively this aspect of PCB analysis whether it be for water, soils, sediments, biological mediums or others. Some books, such as Erickson's "Analytical Chemistry of PCB",[1] provide an excellent synthesis of the question.

Different kinds of apparatus are available for the liquid-liquid or solid-liquid extraction process.

- Liquid-liquid extractor: this device, which is perfectly suited for the extraction of PCBs in water, is used with extraction solvents that are heavier than water (i.e., CH_2Cl_2, $CHCl_3$) so that the solvent loaded with PCBs can be recycled;

- solid-liquid extractor: a Soxhlet device is perfectly suited for the extraction of PCBs in solid samples such as soils and sediments and can be used with a wide variety of solvents.

In addition to this conventional equipment, mention should be made of what is known as the Kuderna-Danish device used to make samples more concentrated (solvent/PCB).

All of this equipment can be dismantled and cleaned. However, since these time-consuming operations lengthen the whole analysis process, they should be reserved for the more difficult cases. For routine extraction and concentration of PCBs, simpler operations such as mechanical shaking or concentration by selective evaporation should suffice.

It should be noted that solid-liquid extraction is tending to replace liquid-liquid extraction. For example, ready-to-use cartridges, usually made of grafted silica, have two different functions:

- Samples purification through the elution of the PCBs and the retention of certain impurities, in the conventional cleanup operations, florisil performs this task;

- extraction of PCBs (e.g., from types of water); the cartridge, which can be used to concentrate the PCBs simply through the elution of a large volume, will then desorb them when a suitable solvent is applied.

ANALYSIS OF PCBs IN WATER

Extraction and preparation of samples

The extraction and cleanup of samples was dealt with earlier. As was indicated, this can be done very simply with excellent results. For example, the PCBs can be extracted by being vigorously shaken mechanically with dichloromethane or petroleum ether as solvent and then placed in an ultrasonic bath for a short while. After decantation, the organic phase is concentrated by selected evaporation.

Many liquid-liquid extraction solvents are available: dichloromethane, ether, petroleum ether, chloroform, etc. (or mixtures of these solvents).

46

Direct evaluation

Deactivated capillary chromatographic columns can now be used for injecting organically polluted water. Delicate liquid-liquid extraction preparation can thus be avoided, although it is undoubtedly necessary to clean up the water on a suitable cartridge and possibly increase sample concentration. This method should be compared from the standpoint of convenience and accuracy with the solid-liquid extraction method on cartridges followed by elution with an adequate solvent.

Sensitivity and accuracy

The method can be very sensitive but a large volume of water must be used. It is convenient with extraction-concentration cartridges but much less so for liquid-liquid exraction. However, a PCB content in the water of the order of 0.1 ppb (0.1 g/l) can be easily evaluated at the present time.

The accuracy, which the author considers to be limited (i.e., 10 to 20% at best), is nevertheless adequate for the interpretation to be made.

ANALYSIS OF PCBs IN SOILS

Extraction and preparation of samples

Solid-liquid extraction can be performed with Soxhlet equipment, but simpler processes with equally convincing results are available. These include mechanical shaking of a mixture consisting of correctly dried and pulverized soil and about 10 times as much solvent, and conventional operations involving cleanup and possibly concentration.

Commonly-used solvents include methanol, chloroform, hexane, petroleum ether, a hexane-acetone mixture, and an ethyl-dichloromethane acetate mixture.

Sensitivity and accuracy

The remarks made for determining PCBs in water also apply here, however the cleanup must be more rigorous since many substances can be extracted from the soil. Sensitivity can thus be adversely affected but still remain well within the standards set forth in the applicable regulations.

CONTAMINATION OF OTHER MATERIALS

Still in the field of electrotechnical applications, incidents may occur following which the PCB content of concrete may also have to be evaluated[12]. The concrete is therefore pulverized and treated in the same way as soil samples.

The contamination of equipment such as transformers either to be recycled or simply to be removed empty can also be cause for concern. After the equipment has been cleaned, the solvent used should be monitored. For example, it may be necessary to evaluate the PCB content in perchloroethylene, which, as is the case for any other solvents, poses no problems.

CONCLUSION

The analysis of PCBs in various mediums, such as water, soils or oil, is a delicate operation that has been mastered. Gas chromatography with an electron capture detector, which can be coupled in problem cases with a mass spectrometer, is the most suitable method. This article briefly outlined the calculation methods available. The excellent performance of capillary columns call into question current testing standards, which warrant review and revision. Furthermore, the traditional sample preparation methods (i.e., PCB extraction, cleanup, concentration) are evolving and being simplified considerably with the introduction of a wide variety of very specific ready-to-use cartridges. Nevertheless, many tests remain to be performed in this area by laboratories already equipped for PCB analysis so that the effectiveness of their work can be validated.

REFERENCES

1. M.D. Erickson, "Analytical Chemistry of PCB," Butterworths (1986).
2. J. Castonguay, Proc. of IEEE Montech 86 Conference, (Montréal), pp. 56-60 (1986).
3. B.M. Hughes, Proc. of IEEE Montech 86 Conference, (Montréal), pp. 46-50 (1986).
4. ASTM test D 4059 (1986), "Analysis of Polychlorinated Biphenyls in Insulating Liquids by Gas Chromatography".
5. IEC Publication 588-2 (1978), "Askarels for Transformers and Capacitors-Testing Methods".
6. P. Mattson and S. Nygren, J. Chromatography $\underline{124}$, 265 (1976).
7. R. Gordon, J. Szita and E.J. Faeder, Anal. Chem. $\underline{54}$, 478 (1982).
8. See "PCB Analysis by Packed Column and Capillary Column" in Bulletin GC 817 (1984).
9. M. Buser and C. Rappe, Anal. Chem. $\underline{56}$, 442 (1984).
10. F.L. Shore, T.W. Campbell, D.A. Hayes, E.D. Hardin and V.A. Fishman, Proc. of IEEE Montech 86 Conference (Montréal), pp. 64-67 (1986).
11. For example, the "Clor-N-Oil" kit manufactured by Dexsil Corp. and the "Kwik-Skrene Analytical Testing System" manufactured by Syprotec Inc.
12. See J.P. Woodyard and E.M. Zoratto, "State-of-the-Art Technology for PCB Decontamination of Concrete," in this book.

HUMAN HEALTH HAZARDS

POLYCHLORINATED BIPHENYLS – HUMAN HEALTH EFFECTS

S. Safe

Department of Veterinary Physiology and Pharmacology
Texas A&M University
College Station, TX 77843

INTRODUCTION

Commercial polychlorinated biphenyls (PCBs) are formulations prepared by the chlorination of biphenyl.[1-4] The chlorine content of the products are directly dependent on the relative amount of Cl_2 used in the reaction process and commercial formulations with variable chlorine content have been produced and marketed. Most of the producers market PCB formulations with a variable chlorine content, for example, Aroclor 1221 and 1260, two commercial PCBs produced by the Monsanto Chemical Corporation contain 21 and 60% chlorine by weight. The degree of biphenyl chlorination significantly alters the properties of these industrial mixtures and this variability accounts for their widespread applications. The highly versatile PCBs have enjoyed a diverse use pattern which takes advantage of their wide range of physical properties, their chemical stability and miscibility with organic compounds. These characteristics have resulted in the use of PCBs as hydraulic fluids, plasticizers, adhesives, heat transfer fluids, wax extenders, dedusting agents, organic dilutents/extenders, lubricants, flame retardants, and as dielectric fluids in capacitors and transformers. The detection of PCBs in the environment resulted in a voluntary restriction ban on all "open" uses of these compounds whereas their use as dielectric fluids ("closed") was permitted until the late 1970s. Although total world production figures for PCBs are not readily available, it has been estimated that approximately 1.4×10^9 lbs was produced in the U.S. during the period 1930 to 1975.

In 1966, PCBs were first identified as environmental contaminants during the analysis of environmental extracts for DDT and related metabolites.[5] It is now apparent that PCBs are among the widespread pollutants in the global ecosystem and have been identified in the air, water, sediments, fish and wildlife, domestic animals and human adipose tissue, blood and milk.[6-28]

PCBS – BIOLOGIC AND TOXIC EFFECTS

The toxic effects elicited by PCBs have been extensively investigated and reviewed[29-42] and the following responses have been reported:

(a) acute lethality at relative high dose levels (> 1000 mg/kg) in most animal species;

(b) body weight loss or a wasting syndrome;

(c) thymic and splenic atrophy and impaired immune function;

(d) teratogenicity and reproductive problems;

(e) dermal toxicity including chloracne;

(f) a role in modulating carcinogenesis and carcinogenicity;

(g) hepatotoxic effects including porphyria;

(h) the induction of diverse enzyme systems including several hepatic drug-metabolizing enzymes (e.g. cytochrome P-450 dependent monooxygenases).

The occurrence and severity of the toxic symptoms noted above are species, strain, sex and age-dependent and not all the responses, with the possible exception of enzyme induction, are observed in any single animal species. It is also apparent that other halogenated aromatics, including the polychlorinated dibenzo-p-dioxins (PCDDs) and dibenzofurans (PCDFs) elicit comparable species-dependent biologic and toxic effects and it has been proposed that these toxins act through a comparable mechanism which involves initial binding to the cytosolic aryl hydrocarbon (Ah) receptor protein present in target tissues.[33,40-42]

One of the hallmarks of receptor-mediated responses is the stereoselective interaction of ligands with their respective receptors and therefore structure-activity relationships (SARs) are used as evidence to support proposed receptor-mediated processes. The SAR for PCBs have been extensively investigated and provide strong evidence for the role of the Ah receptor in initiating the toxic syndromes caused by PCBs and related compounds.

The most toxic PCB congeners, namely 3,3',4,4'-tetra-, 3,3',4,4',5-penta- and 3,3',4,4',5,5'-hexachlorobiphenyl can assume coplanar conformations and are approximate isostereomers of the highly toxic hydrocarbon, 2,3,7,8-tetrachlorodibenzo-p-dioxin (TCDD). These three PCBs are potent AHH inducers[41-47], elicit biologic and toxic effects comparable to those reported for 2,3,7,8-TCDD[29-42], exhibit high affinities for the Ah receptor[48], and the results support a common mechanism of action for the toxic PCBs and PCDDs. Based on PCB structure-activity relationships, the most active compounds are substituted at the para and at least one meta position of both phenyl rings and do not contain any ortho-chloro substituents.[41-47] These SARs define a subset of 4 compounds, namely 3,4,4',5-tetra-, 3,3',4,4'-tetra-, 3,3',4,4',5-penta- and 3,3',4,4',5,5'-hexachlorobiphenyl that are highly toxic but are present as only trace components in the commercial PCB mixtures.[49] It is conceivable that the observed biologic and toxic effects of PCB formulations may either be due to the trace levels of the coplanar PCBs, and/or the presence of highly toxic PCDF impurities[50] or the presence of other toxic PCBs which have not been defined by the proposed structure-activity correlations.[43,44] Studies in our laboratory with Aroclor 1254 showed that after Florisil column chromatography, the cleaned-up (PCDF-free) commercial PCBs and the crude mixture exhibited comparable AHH induction potencies. This suggested that unidentified toxic PCBs may be present in the commercial mixtures and a more comprehensive PCB structure-activity study was undertaken to resolve this problem.[41,42,45-47]

The introduction of a single ortho-chloro substituent into the biphenyl ring results in decreased coplanarity between the two phenyl rings due to steric interactions between the bulky ortho-chloro and hydrogen substituents. The effects of ortho substituents on PCB activity were tested by synthesizing all the mono-ortho analogs of the most active coplanar PCBs (i.e., 3,4,4',5-tetra-, 3,3,4,4-tetra-, 3,3',4,4',5-penta- and 3,3',4,4',5,5'-hexachlorobiphenyl) and determining the mixed-function oxidase enzyme-inducing activities in immature male Wistar and Long-Evans rats.[45-47] All of these compounds induce AHH and DMAP N-demethylase in the Wistar rats and the related cytochromes P-450a-P-450e in the Long-Evans rats. It was apparent that the mono-ortho analogs of the coplanar PCBs resembled phenobarbital (PB) plus 3-methylcholanthrene (MC) (coadministered) and Aroclor 1254 in their mode of drug-metabolizing enzyme induction. A comparison of the coplanar and mono-ortho coplanar PCBs clearly shows that the ortho-chloro substituent diminishes but does not eliminate Ah receptor binding affinities[48] of the mono-ortho coplanar PCBs and this is also accompanied by decreased AHH/EROD induction potencies of these compounds. These results identify the structures of the active chlorinated biphenyls components in the commercial PCBs since several mono-ortho coplanar PCBs, including 2,3,3',4,4'-penta-, 2,3',4,4',5-penta-, 2,3,3',4,4',5-hexa- and 2,3,3',4,4',5,5'-heptachlorobiphenyl have been identified in commercial formulations.[1,51-]

Moreover, although the toxicities of the mono-ortho coplanar PCBs have not been systematically investigated, many of these compounds elicit toxic effects similar to 2,3,7,8-TCDD and related halogenated aromatics.[41,45,54-56] It is conceivable that some diortho-chloro substituted PCBs may also exhibit some 2,3,7,8-TCDD-like activity; however, their relative potencies would be much less than the coplanar or monoortho coplanar PCBs.

PCBS - HUMAN HEALTH EFFECTS

Any assessment of adverse human health effects of PCBs must consider the following factors which may influence the potential toxicity of these compounds, namely: (i) the route(s) of exposure, (ii) duration of exposure, (iii) the composition of the commercial PCB products (e.g., degree of chlorination), and (iv) the presence and levels of potentially toxic polychlorinated dibenzofurans. Due to these variables, it is not surprising that there are significant differences in the effects of PCBs on different exposed groups.

Yusho and Yu-Cheng Poisoning

In 1968, a mass food poisoning was reported in the Fukuoka and Nagasaki prefectures in southwestern Japan where over 1600 people were poisoned after consuming rice oil contaminated with a commercial PCB industrial fluid, Kanechlor 400.[57-60] In 1979, a similar poisoning (i.e. Yu-Cheng) incident occurred in Taiwan in which over 2000 people were poisoned after consuming PCB-contaminated rice oil.[61,62] The most frequently reported initial symptom of this toxicosis was chloracne and related dermal problems; in addition, a broad spectrum of effects were reported and these are typified by the toxic symptoms observed by Kuratsune and coworkers (Table 1).[58,59,62] Moreover, many of these same symptoms were also observed in Yu-Cheng patients[63] and it is clear from numerous studies that both poisonings share a common etiology. Kuratsune and coworkers[58] have also demonstrated the dose-response relationship between the consumption of Kanechlor-contaminated rice oil and the severity of Yusho poisoning (Table 2). The severity of the acute poisoning effects of Yusho victims have been monitored since the accident[64,65] and between 1969-1975, there was significant recovery from the mucocutaneous lesions in 64% of the patients. However, it was also reported that other symptoms such as headaches and stomach aches, numbness of the extremities, coughing and bronchial disorders, and joint pains were common in many of these patients. Children poisoned in the Yusho incident exhibited retarded growth, abnormal tooth development and

Table 1. Percent Distribution of Signs and Symptoms of Yusho and Yu-Cheng Poisoning[59,63]

Symptoms	Males (Yusho) (n=89)	Males* (Yu-Cheng) (n=15)	Females (Yusho) (n=100)	Females* (Yu-Cheng) (n=12)
Dark brown pigmentation of nails	83.1	86.6	75.0	83.3
Distinctive hair follicles	64.0	40	56.0	41.6
Increased sweating at palms	50.6		55.0	
Acne-like skin eruptions	87.6	86.6	82.0	83.3
Red plaques on limbs	20.2		16.0	
Itching	42.7		52.0	
Pigmentation of skin	75.3		72.0	
Swelling of limbs	20.2		41.0	
Stiffened soles in feet and palms of hands	24.7	46.6	29.0	25
Pigmented mucous membrane	56.2		47.0	
Increased eye discharge	88.8	93.3	83.0	91.6
Hyperemia of conjunctiva	70.8	66.6	71.0	75
Transient visual disturbance	56.2		55.0	
Jaundice	11.2		11.0	
Swelling of upper eyelids	71.9	86.6	74.0	91.6
Feeling of weakness	58.4		52.0	
Numbness in limbs	32.6	53.3	39.0	56
Fever	16.9		19.0	
Hearing difficulties	18.0		19.0	
Spasm of limbs	7.9		8.0	
Headache	30.3		39.0	
Vomiting	23.6		28.0	
Diarrhea	19.1		17.0	

*Pigmentation of lips, black color of nose, pigmentation of conjunctivae, hypesthesia, deformity of nails, pigmentation of gingivae, amblyopia were also observed

Table 2. Relationship Between the Amount of Kanechlor-Contaminated Rice Oil Consumed and Clinical Severity of Yusho[58]

Estimated Amount of Oil Consumed (ml)	Nonaffected		Light Cases		Severe Cases	
	No.	%	No.	%	No.	%
< 720	10	12	39	49	31	39
720 - 1440	0	0	14	31	31	69
> 1440	0	0	3	14	18	86

newborns exhibited systemic pigmentation and were undersized. The changes in the severity of the dermal toxicity of Yusho oil in a group of patients 2, 7 and 12 years after the original outbreak of Yusho poisoning are summarized in Table 3 and indicate that there is a gradual recovery from the severe skin problems.[64,65]

Table 3. Distribution of Yusho Patients According to Skin Severity Grades in 1971, 1976 and 1981[64,65]

Skin severity[a] Index	Case Numbers		
	1971	1976	1981
Ø	4	25	56
I	49	27	22
II	32	14	14
III	31	20	13
IV	13	4	2
Total	129	90	107

[a]Grade Ø: No skin eruption.
Grade I: Circumscribed pigmentation of skin.
Grade II: Black comedones.
Grade III: Acneiform eruptions.
Grade IV: Extensive distribution of the acneiform eruptions.

The severe toxicoses observed in Yusho patients exposed to dietary levels of PCBs were not seen in occupationally exposed workers with comparable or higher serum PCB levels. For example, the mean PCB blood levels of Yu-Cheng victims taken a short time after the accident varied from 39-101.7 ppb[61], whereas the PCB serum levels in occupationally-exposed workers can be much higher. Several studies have shown that trace levels of PCDFs contaminate most commercial PCBs[66-75] and Table 4 summarizes the relative concentrations of PCDFs in a commercial PCB (Kanechlor 400) contaminated Yusho oil and Yu-Cheng oil. The ratio of PCDFs/PCBs in the Yu-Cheng and Yusho oils was 2.4 x 10^{-3} and 9.0 x 10^{-3} respectively, whereas the ratio in Kanechlor 400 was 3.3 x 10^{-5} thus illustrating the relatively high levels of the PCDFs in the toxic contaminated rice oils. Several recent studies have focused on the identification and quantitation of PCBs and PCDFs in Yusho and Yu-Cheng patients[76-83] and Table 5 summarizes the serum levels of these compounds in several exposed groups and a control Japanese population. It was

Table 4. Concentrations of PCBs, PCQs and PCDFs and their ratios in various materials[76]

Sample	PCBs, ppm	PCQs, ppm	PCDFs, ppm	PCQs/PCBs, g/g	PCDFs/PCBs, g/g
Yu-Cheng Oil					
Y-1a	51	10	0.14	2.0×10^{-1}	2.7×10^{-3}
Y-2b	54	18	0.10	3.3×10^{-1}	1.9×10^{-3}
Y-3b	69	24	0.18	3.5×10^{-1}	2.6×10^{-3}
Y-4a	22	9	—c	4.1×10^{-1}	—
Y-5b	113	38	—c	3.4×10^{-1}	—
Average	62	20	0.14	3.3×10^{-1}	2.4×10^{-3}
Yusho Oil					
Feb. 5, 1968 (prod. date)	968	866	7.40	8.9×10^{-1}	7.6×10^{-3}
Feb. 9, 1968 (prod. date)	151	490	1.90	3.2	1.3×10^{-2}
Feb. 10, 1968 (prod. date)	155	536	2.25	3.5	1.4×10^{-2}
Average	430	630	3.85	1.5	9.0×10^{-3}
Kanechlor 400	999,800	209	33	2.1×10^{-4}	3.3×10^{-5}

a,bSamples collected from a school cafeteria and victims' homes, respectively

cNo analysis

apparent that the PCB levels in the recently exposed Yu-Cheng patients were higher than the levels in Yusho oil victims, one group of occupationally exposed workers and the control (Japanese population). However, the levels were not significantly greater than those observed in a group of capacitor workers whose last exposure was 9 years prior to the analysis. It was evident from the analytical data that the persistence of the PCDFs in the Yu-Cheng population was the major difference between this group and all the others, moreover analysis of the PCDFs in the exposed individuals[78] has shown that among the most persistent congeners were the highly toxic 2,3,7,8-tetra-, 1,2,3,7,8-penta-, 2,3,4,7,8-penta- and 1,2,3,4,7,8-hexachlorodibenzofuran.

Kunita and coworkers[79] have reported the relative toxicity of purified PCBs and PCDFs obtained by fractionation of a KC-400 PCB preparation used as a heat exchanger. Table 6 summarizes the relative toxicities of the PCDFs (III) and PCBs (II) and a "reconstituted mixture"

Table 5. PCB, PCQ, and PCDF Levels in the Blood of Taiwanese and Japanese Poisoned Patients, Workers Occupationally Exposed to PCBs, Health Persons, and in Toxic Rice Oils[74,79]

Sample	No.	Degree of severity of clinical signs	Concentration (mean ± SD, ppb)			PCBs:PCQs:PCDFs
			PCBs	PCQs	PCDFs	
	5	None	12 ± 6	1.7 ± 1.1	0.024 ± 0.018	100:14:0.15
	24	Slight	33 ± 13	7.9 ± 3.7	0.062 ± 0.024	100:21:0.16
Taiwanese Patients	14 1	Moderate	43 ± 11	8.2 ± 3.5	0.079 ± 0.030	100:19:0.19
	24	Heavy	49 ± 20	11.0 ± 5.2	0.100 ± 0.040	100:23:0.20
	67		42 ± 17	8.6 ± 4.8	0.076 ± 0.038	100:20:0.18
Japanese consumers of Yusho oil	56 11		6 ± 4	2.0 ± 2.0	ND	100:32:<0.17
Japanese worker occupationally exposed to fresh PCB	69 9		45 ± 49	ND	---	100:<0.04:—
Japanese worker occupationally exposed to used PCB	3 9		19 ± 11	0.9 ± 0.9	---	100:5:—
Japanese healthy subjects	60		2	ND	---	100:<1:—

Table 6. Etiology of Yusho Disease - Role of PCDFs Toxicology of Yusho Oil Fractions

Fraction	Dose (mg/rat/day)	Thymus Weight	Body Weight Gain (g)
I Control	--	0.29	255
II PCBs	1.0	0.27	265
III PCDFs	0.01	0.12	165
IV II + III*	1.0 + 0.01	0.09	140

*Resembles composition of Yusho oil and contains PCQs (Amer. J. Ind. Med. 5, 45, 1984)

(IV) which resembles Yusho oil in rats. The results (Table 6) clearly
show that PCDFs are the toxic component of the Yusho mixture.

A recent study in my laboratory[85] also confirmed that the PCDFs were
the major etiologic agents in Yusho oil. Two reconstituted mixtures
containing 5 PCDFs (2,3,7,8-tetra-, 1,2,4,7,8-penta-, 1,2,3,7,8-penta-,
2,3,4,7,8-penta- and 1,2,3,4,7,8-hexachlorodibenzofurans; 7.4, 6.1, 19.0,
29.4 and 38.1% respectively) and 6 PCBs (2,3',4,4',5-penta-,
2,2',4,4',5,5'-hexa-, 2,2',3,4,4',5'-hexa-, 2,3,3',4,4',5-hexa-,
2,2',3,4,4',5,5'-hepta- and 2,2',3,3',4,4',5-heptachlorobiphenyls; 5.7,
22.6, 28.2, 12.3, 19.1 and 12.2% respectively) were prepared from pure
synthetic standards. These compounds and their relative concentrations
were similar to those identified in Yusho victims. The mixtures were
administered in a dose-response fashion to immature male Wistar rats and
the biologic and toxic effects measured in this study [i.e., body weight
loss, thymic atrophy and the induction of hepatic microsomal aryl
hydrocarbon hydroxylase (AHH)] are typically caused by the toxic PCDFs,
PCBs and related halogenated aromatics[33,41,86]; the results were used to
derive the relative potencies of the two mixtures. The data summarized
in Table 7 show that the PCDFs are from 680 to 2210 times more toxic than
the PCBs and since the ratios of PCBs:PCDFs in Yusho/Yu-Cheng victims and
in the contaminated rice oils are generally < 500:1, these results also
suggest that the PCDFs are the major etiologic agents in these accidental
poisonings.

Table 7. Etiology of Yusho Disease - Role of PCDFs Comparative
Toxicology: PCDFs vs. PCBs

Reconst.* Mixture	Relative Potencies		
	AHH Induct.	Body Weight Loss	Thymic Atrophy
PCBs	1	1	1
PCDFs	680	900	2210

*Compounds identified in Yusho victims liver (PCB/PCDF 5/1)
(Chemosphere 13, 507, 1984)

Occupational Exposure to PCBs

Workplace exposure to commercial PCBs can result in significant
uptake of these compounds and this is reflected in the high serum or
adipose tissue levels detected in groups of exposed workers. The effects
of occupational exposure to PCBs are comparable to some of the effects

observed in laboratory animals and in Yusho poisoning. Chloracne and related skin problems have been observed in several groups of workers[87-94] and it was suggested that the air concentrations of commercial PCBs > 0.1 mg/m[3] were associated with this effect.[94] It was also reported that after occupational exposure to PCBs was terminated, there was a gradual decrease in the severity and number of dermatological problems in the exposed workers and this paralleled a decrease in their serum levels of PCBs.[90]

The effects of occupational PCB exposure to PCB on the levels of several serum clinical chemical and hematological parameters have been reported by several groups.[88,89,91,95-98] Mildly elevated SGOT and - glutamyl transpeptidase (GGTP) suggest some liver damage and induction of hepatic monooxygenase enzymes. These results are not surprising in light of animal studies which report that PCBs frequently cause hepatomegaly accompanied by monooxygenase enzyme induction.[92] Alvares and coworkers[93] have reported that although the serum enzymes were not elevated in 5 workers exposed to PCBs, their antipyrine half-lives were decreased and this was consistent with induced hepatic monooxygenases. Warshaw and colleagues reported a relatively high incidence of pulmonary dysfunction in capacitor workers[94] and the symptoms included coughing, wheezing, tightness in the chest, and upper respiratory problems or eye irritation.

Retrospective mortality studies[99] in 2,567 workers (> 3 months employment) from two capacitor manufacturing plants showed that mortality in both plants was lower than expected and there were no significant increases in either liver or rectal cancer. A study on 31 workers exposed to Aroclor 1254 and other chemicals while employed in a New Jersey petrochemical plant[100] showed an increased incidence of malignant melanomas (0.04% expected; 0.13% observed); however, due to the small populations involved in this survey, more comprehensive long term epidemiologic studies will be required to fully assess the carcinogenicity of PCBs. However, it is apparent from most reports that occupational exposure to PCBs results in characteristic toxic symptoms which appear to be reversible.

Environmental Exposure to PCBs

Serum levels of PCBs in the general population are usually < 10-15 ppb and these concentrations are significantly lower than observed in occupationally exposed workers in which serum PCB concentrations as high as 3330 ppb have been reported.[96] Since industrial exposure to high levels of PCBs elicit toxic symptoms which are reversible, it is unlikely that environmental uptake of PCB by the general population results in significant adverse human health effects. Individuals who consume relatively large amounts of fish represent a small subpopulation exposed to higher levels of PCBs and this is reflected in elevated serum PCB concentrations.[101-103] The mean serum PCB levels in one such group from Triana, Alabama were 17.2 ppb; moreover, there was a correlation between serum PCB concentrations and elevated blood pressure. The incidence of borderline and definite hypertension was increased 30% over the expected values for this population. It was noted that "the colinearity of DDT and PCB serum concentrations in this rural population, exposed to both chemical families through consumption of contaminated fish, precludes any certainty regarding which family of chlorinated hydrocarbons may be correlated with blood pressure."[103]

Recent studies in our laboratory[104,105] have investigated the interactive effects of relatively high (20-750 umol/kg) but non-toxic doses of Aroclor 1254 with toxic doses of 2,3,7,8-TCDD administered to C57BL/6J mice. The experiments were designed to determine whether Aroclor 1254, a weak Ah receptor agonist could antagonize the toxic effects of 2,3,7,8-TCDD. The results demonstrated that at specific molar ratios of Aroclor 1254/2,3,7,8-TCDD, there was significant antagonism by Aroclor 1254 of the teratogenic (ratio: ~ 12000/1) and immunotoxic (ratio: 1,340 to 20,160/1) effects caused by 2,3,7,8-TCDD. Moreover, the ratios of antagonist/agonist in which antagonism was observed overlapped with the ratios of PCBs/toxic PCDDs plus PCDFs observed in environmental samples and the general population.[106,107] These data show that at least in laboratory animals PCBs possess some beneficial activity since they act as 2,3,7,8-TCDD antagonists.

Future studies on the adverse human health effects of PCBs will require more standardized methods of congener specific PCB analysis, the development of more sensitive techniques to determine subtle toxic effects caused by relatively low exposures to PCBs (e.g. environmental) and a more thorough knowledge of the interactive effects of PCBs with other environmental pollutants. The most appropriate indicators of the adverse health effects of PCBs and their role in carcinogenicity will be determined from the continuing epidemiological and retrospective studies on occupationally exposed workers.

ACKNOWLEDGEMENTS

The financial assistance of the Texas Agricultural Experiment Station, the Chester Reed Endowment, the National Institutes of Health, and the Environmental Protection Agency are gratefully acknowledged.

REFERENCES

1. O. Hutzinger, S. Safe, and V. Zitko, "The Chemistry of PCBs," CRC Press, Boca Raton (1974).

2. U.A.Th. Brinkman and A. De Kok, Production properties and uses, in: "Halogenated Biphenyls, Terphenyls, Naphthalenes, Dibenzodioxins and Related Products," R.D. Kimbrough, ed., Elsevier/North Holland, Amsterdam (1980).

3. I. Pomerantz, J. Burke, D. Firestone, J. McKinney, J. Roach, and J. Trotter, Chemistry of PCBs and PBBs, Environ. Health Perspect. 24:133 (1978).

4. C. Rappe and H.R. Buser, Chemical properties and analytical methods, in: "Halogenated Biphenyls, Terphenyls, Naphthalenes, Dibenzodioxins and Related Products," R.D. Kimbrough, ed., Elsevier/North Holland, Amsterdam (1980).

5. Anon., Report of a new chemical hazard, New Sci., 32:621 (1966).

6. R.W. Risebrough, P. Rieche, S.G. Herman, D.B. Peakall, and M.N. Kirven, Polychlorinated biphenyls in the global ecosystem, Nature (London) 220:1098 (1968).

7. L. Fishbein, Chromatographic and biological aspects of polychlorinated biphenyls, J. Chromatogr. 68:345 (1972).

8. K. Ballschmiter, H. Buchert, and S. Bihler, Baseline studies of the global pollution, Fres. Z. Anal. Chem. 306:323 (1981).

9. K. Ballschmiter, Ch. Scholz, H. Buchert, M. Zell, K. Figge, K. Polzhofer, and H. Hoerschelmann, Studies of the global baseline pollution, Fres. Z. Anal. Chem. 309:1 (1981).

10. F.W. Kutz, S.C. Strassman, and J.F. Sperling, Survey of selected organochlorine pesticides in the general population of the United States: fiscal years 1970-1975, Ann. N.Y. Acad. Sci. 320:60 (1979).

11. K. Fujiwara, Environmental and food contamination with PCBs in Japan, Sci. Total Environ. 4:219 (1975).

12. M. Wasserman, D. Wasserman, S. Cucos, and H.J. Miller, World PCBs map: storage and effects in man and his biologic environment in the 1970s, Ann. N.Y. Acad. Sci. 320:69 (1979).

13. K. Higuchi, Ed., "PCB Poisoning and Pollution," Kodansha, Tokyo (1976).

14. D. Mackay, S. Paterson, S.J. Elsenreich, and M.S. Simmons, Eds., "Physical Behavior of PCBs in the Great Lakes," Ann Arbor Science, Ann Arbor (1983).

15. G.R. Harvey and W.G. Steinhauer, Atmospheric transport of polychlorobiphenyls to the North Atlantic, Atmos. Environ. 8:777 (1974).

16. S. Tanabe, H. Hidaka, and R. Tatsukawa, PCBs and chlorinated hydrocarbon pesticides in Antarctic atmosphere and hydrosphere, Chemosphere 12:277 (1983).

17. E. Atlas and C.S. Giam, Global transport of organic pollutants: ambient concentrations in the remote marine atmospheres, Science 211:163 (1981).

18. C.L. Stratton and J.B. Sosebee, Jr., PCB and PCT contamination of the environment near sites of manufacture and use, Bull. Environ. Contam. Toxicol. 10:1229 (1976).

19. T.J. Murphy, J.C. Pokojowczyk, and M.D. Mullin, Vapor exchange of PCBs with Lake Michigan: the atmosphere as a sink for PCBs, in: "Physical Behavior of PCBs in the Great Lakes," D. Mackay, S. Paterson, S.J. Eisenreich and M.S. Simmons, eds., Ann Arbor Science, Ann Arbor (1983).

20. International Joint Commission, IJC Great Lakes water quality—Appendix E, Status Report on the Persistent Toxic Pollutants in the Lake Ontario Basin (1977).

21. P.P. Kauss, K. Suns, and E.H. Buckley, Monitoring of PCBs in water, sediments and biota of the Great Lakes--some recent examples, in: "Physical Behavior of PCBs in the Great Lakes," D. Mackay, S. Paterson, S.J. Eisenreich, and M.S. Simmons, eds., Ann Arbor Science, Ann Arbor (1983).

22. J.R. Sullivan, J. Delfino, C.R. Buelow, and T.B. Sheffy, Polychlorinated biphenyls in the fish and sediment of the Lower Fox River, Wisconsin, Bull. Environ. Contam. Toxicol. 30:58 (1983).

23. J. Passavirta, J. Sarkka, K. Surma-Aho, T. Humpi, T. Kuokkanen, and M. Marttinen, Food chain enrichment of organochlorine compounds and mercury in clean and polluted lakes of Finland, Chemosphere 12:239 (1983).

24. M.E. Zabik, C. Merrill, and M.J. Zabik, PCBs and other xenobiotics in raw and cooked carp, Bull. Environ. Contam. Toxicol. 28:710 (1982).

25. D.L. Stalling, J.N. Huckens, J.D. Petty, J.L. Johnson, and H.O. Sanders, An expanded approach to the study and measurement of PCBs and selected planar aromatic environmental pollutants, Ann. N.Y. Acad. Sci. 320:48 (1979).

26. H. Brunn and D. Manz, Contamination of native fish stock by hexachlorobenzene and polychlorinated biphenyl residues, Bull. Environ. Contam. Toxicol. 28:599 (1982).

27. K. Wickstrom, H. Pyysalo, and M. Pettila, Organochlorine compounds in the liver of cod (Gadus Morhua) in the northern Baltic, Chemosphere 10:999 (1981).

28. P. Olsen, H. Settle, and R. Swift, Organochlorine residues in wings of ducks in south-eastern Australia, Aust. Wildl. Res. 7:139 (1980).

29. E.E. McConnell, Acute and chronic toxicity, carcinogenesis, reproduction teratogenesis and mutagenesis in animals, in: "Halogenated Biphenyls, Terphenyls, Naphthalenes, Dibenzodioxins and Related Products," R.D. Kimbrough, ed., Elsevier/North Holland, Amsterdam (1980).

30. L.H. Garthoff, L. Friedman, T.M. Farber, K.K. Locke, T.J. Sobotka, S. Green, N.E. Hurley, E.L. Peters, G.E. Story, F.M. Moreland, C.H. Graham, J.E. Keys, M.J. Taylor, J.V. Scalera, J.E. Rothlein, E.M. Marks, F.E. Cerra, S.B. Rodi, and E.M. Sporn, Biochemical and cytogenetic effects in rats caused by short-term ingestion of Aroclor 1254 or FireMaster BP-6, J. Toxicol. Environ. Health 3:679 (1977).

31. R. Kimbrough, J. Buckley, L. Fishbein, G. Flamm, L. Kasza, W. Marcus, S. Shibko, and R. Teske, Animal toxicology, Environ. Health Perspect. 24: 173 (1978).

32. A. Matthews, G. Fries, A. Gardner, L. Garthoff, J. Goldstein, Y. Ku, and J. Moore, Metabolism and biochemical toxicity of PCBs and PBBs, Environ. Health Perspect. 24:147 (1978).

33. A. Poland and J.C. Knutson, 2,3,7,8-Tetrachlorodibenzo-p-dioxin and related halogenated aromatic hydrocarbons: examination of the mechanism of toxicity, Ann. Rev. Pharmacol. Toxicol. 22:517 (1982).

34. F.J. Dicarlo, J. Steiffer, and V.J. Decarlo, "Assessment of the hazards of polybrominated biphenyls," Tech. Rep. No. EPA-560/6-77-037. U.S. Environmental Protection Agency, Washington, D.C.

35. A. Parkinson and S. Safe, Aryl hydrocarbon hydroxylase induction and its relationship to the toxicity of halogenated aryl hydrocarbons, Toxicol. Environ. Chem. Rev. 4:1 (1981).

36. R.D. Kimbrough, The toxicity of polychlorinated polycyclic compounds and related compounds, Crit. Rev. Toxicol. 2:445 (1974).

37. L. Fischbein, Toxicity of chlorinated biphenyls, Annu. Rev. Pharmacol. 14:139 (1974).

38. E.E. McConnell and J.A. Moore, Toxicopathology characterstics of the halogenated aromatics, Ann. N.Y. Acad. Sci. 320:138 (1979).

39. J.R. Roberts, D.W. Rodgers, J.R. Bailey, and M.A. Rorke, Polychlorinated biphenyls: biological criteria for an assessment of their effects on environmental quality, National Research Council of Canada, NRCC No. 16077, Ottawa, Ontario (1978).

40. S. Safe, S. Bandiera, T. Sawyer, L. Robertson, A. Parkinson, P.E. Thomas, D. Ryan, L.M. Reik, W. Levin, M.A. Denomme, and T. Fujita, PCBs: structure-function relationships and mechanism of action, Environ. Health Perspect. 60:47 (1985).

41. S. Safe, Polychlorinated biphenyls (PCBs) and polybrominated biphenyls (PBBs): biochemistry, toxicology and mechanism of action, CRC Crit. Rev. Toxicol. 13:319 (1984).

42. S. Safe, L.W. Robertson, L. Safe, A. Parkinson, S. Bandiera, T. Sawyer, and M.A. Campbell, Halogenated biphenyls: molecular toxicology, Can. J. Physiol. Pharmacol. 60:1057 (1982).

43. J.A. Goldstein, P. Hickman, H. Bergman, J.D. McKinney, and M.P. Walker, Separation of pure polychlorinated biphenyl isomers into two types of inducers on the basis of induction of cytochrome P-450 or P-448, Chem.-Biol. Interact. 17:69 (1977).

44. A. Poland and E. Glover, Chlorinated biphenyl induction of aryl hydrocarbon hydroxylase activity: a study of structure-activity relationships. Mol. Pharmacol. 13:924 (1977).

45. A. Parkinson, S. Safe, L. Robertson, P.E. Thomas, D.E. Ryan, L.M. Reik, and W. Levin, Immunochemical quantitation of cytochrome P-450 isozymes and epoxide hydrolase in liver microsomes from polychlorinated and polybrominated biphenyls: a study of structure activity relationships, J. Biol. Chem. 258:5967 (1983).

46. A. Parkinson, R. Cockerline, and S. Safe, Induction of both 2-methylcholanthrene- and phenobarbitone-type microsomal enzyme activity by a single polychlorinated biphenyl isomer, Biochem. Pharmacol. 29:259 (1980).

47. A. Parkinson, R. Cockerline, and S. Safe, Polychlorinated biphenyl isomers and congeners as inducers of both 3-methylcholanthrene- and phenobarbitone-type microsomal enzyme activity, Chem.-Biol. Interact. 29:277 (1980).

48. S. Bandiera, S. Safe and A.B. Okey, Binding of polychlorinated biphenyls classified as either PB-, MC- or mixed-type inducers to cytosolic Ah receptor, Chem.-Biol. Interact. 39:259 (1982).

49. L.R. Kamops, W.J. Trotter, S.J. Young, A.C. Smith, J.A.G. Roach, and S.W. Page, Separation and quantitation of 3,3',4,4'-tetrachlorobiphenyl and 3,3',4,4',5,5'-hexachlorobiphenyl in Aroclors using Florisil column chromatography and gas-liquid chromatography, Bull. Environ. Contam. Toxicol. 23:51 (1979).

50. J.G. Vos, J.H. Koeman, H.L. Van Der Mass, M.C. ten Noever de Brauw, and R.H. de Vos, Identification and toxicological evaluation of chlorinated dibenzofuran and chlorinated naphthalene in two commercial polychlorinated biphenyls, Food Cosmet. Toxicol. 18:625 (1970).

51. S. Jensen and G. Sundstrom, Structures and levels of most chlorobiphenyls in the technical PCB products and in human adipose tissue, Ambio 3:70 (1974).

52. K. Ballschmiter and M. Zell, Analysis of polychlorinated biphenyls (PCBs) by glass capillary gas chromatography. Composition of technical Aroclor- and clophen-PCB mixtures, Z. Anal. Chem. 302:20 (1980).

53. S. Safe, L. Safe, and M. Mullin, Polychlorinated biphenyls (PCBs)-congener-specific analysis of a commercial mixture and a human milk extract, J. Agric. Food Chem. 33:24 (1985).

54. H. Yamamoto, H. Yoshimura, M. Fujita, and T. Yamamoto, Metabolic and toxicologic evaluation of 2,3',4,4',5'-pentachlorobiphenyl in rats and mice, Chem. Pharm. Bull. 24:2168 (1976).

55. R.L. Ax and L.G. Hansen, Effects of purified PCB analogs on chicken reproduction, Poult. Sci. 54:895 (1975).

56. S. Yoshihara, K. Kawano, H. Yoshimura, H. Kuroki, and Y. Masuda, Toxicological assessment of highly chlorinated biphenyl congeners retained in the Yusho patients, Chemosphere 8:531 (1979).

57. K. Higuchi (Ed.), "PCB Poisoning and Pollution," Kodansha Ltd. and Academic Press, Tokyo-London-New York (1976).

58. M. Kuratsune, T. Yoshimura, J. Matsuzaka and A. Yamaguchi, Epidemiological study on Yusho, a poisoning caused by ingestion of rice oil contaminated with a commercial brand of polychlorinated biphenyls, Environ. Health Perspect. 1:119 (1972).

59. M. Kuratsune, Yusho, in: "Halogenated Biphenyls, Terphenyls, Naphthalenes, Dibenzodioxins and Related Products," R.D. Kimbrough, ed., Elsevier/North Holland Biomedical Press, Amsterdam (1980).

60. Y. Masuda, H. Kuroki, T. Yamaryo, K. Haraguchi, M. Kuratsune and S.T. Hsu, Comparison of causal agents in Taiwan and Fukuoka PCB poisonings, Chemosphere 11:199 (1982).

61. S-T. Hsu, C-I. Ma, S. Kwo-Hsiung, S.S. Wu, N.H-M. Hsu, C.C. Yeh and S.B. Wu, Discovery and epidemiology of PCB poisoning in Taiwan: a four year followup, Environ. Health Persp. 59:5 (1985).

62. M. Kuratsune and R.E. Shapiro (Ed.), "Poisoning in Japan and Taiwan," Alan R. Liss, Inc., New York (1984).

63. W.M. Li, C.J. Chen and C.K. Wong, PCB poisoning of 27 cases in three generations of a large family, Clinical Med. (Taipei) 7:23 (1981).

64. H. Urabe, H. Koda and M. Asahi, Present state of Yusho patients, Ann. N.Y. Acad. Sci. 320:273 (1979).

65. H. Urabe and M. Asahi, Past and current dermatological status of Yusho patients, Amer. J. Ind. Med. 5:5 (1984).

66. G.W. Bowes, M.J. Mulvihill, B.R.T. Simoneit, A.L. Burlingame and R.W. Risebrough, Identification of chlorinated dibenzofurans in American polychlorinated biphenyls, Nature 256:305 (1975).

67. G.W. Bowes, M.J. Mulvihill, M.R. DeCamp and A.S. Kende, Gas chromatographic characteristics of authentic chlorinated dibenzofurans: identification of two isomers in American and Japanese polychlorinated biphenyls, J. Agr. Food Chem. 23:1222 (1975).

68. M. Morita, J. Nakagawa, K. Akiyama, S. Mimura and N. Isono, Detailed examination of polychlorinated dibenzofurans in PCB preparations and Kanemi Yusho oil, Bull. Environ. Cont. Toxicol. 18:67 (1977).

69. H. Miyata, A. Nakamura and T. Kashimoto, Separation of polychlorodibenzofurans (PCDFs) in Japanese commercial PCBs (Kanechlors) and their heated preparation, J. Food Hyg. Soc. Japan 17:227 (1976).

70. H. Miyata and T. Kashimoto, The finding of polychlorodibenzofurans in commercial PCBs (Aroclor Phenoclor and Clophen) (In Japanese), J. Food Hyg. Soc. Japan 17:434 (1976).

71. H. Miyata, T. Kashimoto and N. Kunita, Detection and determination of polychlorodibenzofurans in normal human tissues and Kanemi rice oils caused "Kanemi Yusho" (In Japanese), J. Food Hyg. Soc. Japan 19:260 (1977).

72. H.R. Buser, C. Rappe and A. Gara, Polychlorinated dibenzofurans (PCDFs) found in Yusho oil and in used Japanese PCB, Chemosphere 7:439 (1978).

73. P.H. Chen, K.T. Chang and Y.D. Lu, Polychlorinated biphenyls and polychlorinated dibenzofurans in the toxic rice-bran oil that caused PCB poisoning in Taichung, Bull. Environ. Contam. Toxicol. 26:489 (1981).

74. T. Kashimoto, H. Miyata, S. Kunita, T.C. Tung, S.T. Hsu, K.J. Chang, S.Y. Tang, G. Ohi, J. Nakagawa and S.I. Yamamoto, Role of polychlorinated dibenzofuran in Yusho (PCB poisoning), Arch. Environ. Health 36:321 (1981).

75. P.H. Chen, Y.D. Lu, M.H. Yang and J.S. Chen, Toxic compounds in the cooking oil which caused PCB poisoning in Taiwan. II. The presence of polychlorinated quaterphenyls and polychlorinated terphenyls (In Chinese), Clin. Med. (Taipei) 7:77 (1981).

76. H. Miyata, S. Fukushima, T. Kashimoto and N. Kunita, PCBs, PCQs and PCDFs in the tissues of Yusho and Yu-Cheng patients, Environ. Health Perspect. 59:67 (1985).

77. T. Kashimoto, H. Miyata, S. Fukushima, N. Kunita, G. Ohi and T-C. Tung, PCBs, PCQs and PCDFs in blood of Yusho and Yu-Cheng patients, Environ. Health Perspect. 59:73 (1985).

78. Y. Masuda, H. Kuroki, K. Haraguchi and J. Nagayama, PCB and PCDF congeners in the blood and tissues of Yusho and Yu-Cheng patients, Environ. Health Perspect. 59:53 (1985).

79. N. Kunita, T. Kashimoto, H. Miyata, S. Fukushima, S. Hall and H. Obana, Causal agents of Yusho, Amer. J. Ind. Med. 5:45 (1984).

80. H. Kuroki and Y. Masuda, Determination of polychlorinated dibenzofuran isomers retained in patients with Yusho, Chemosphere 7:771 (1978).

81. C. Rappe, H.R. Buser, H. Kuroki and Y. Masuda, Identification of polychlorinated dibenzofurans (PCDFs) retained in patients with Yusho, Chemosphere 8:259 (1979).

82. Y. Masuda, H. Kuroki, T. Yamaryo, K. Haraguchi, M. Kuratsune and S.T. Hsu, Comparison of causal agents in Taiwan and Fukuoka PCB poisonings, Chemosphere 11:199 (1982).

83. T. Kashimoto, H. Miyata and N. Kunita, The presence of polychlorinated quaterphenyls in the tissues of Yusho victims, Fd. Cosmet. Toxicol. 19:335 (1981).

84. J. Nagayama, Y. Masuda and M. Kuratsune, Determination of polychlorinated dibenzofurans in tissues of patients with "Yusho," Food Cosmet. Toxicol. 15:195 (1977).

85. S. Bandiera, T. Sawyer, M. Romkes, B. Zmudzka, L. Safe, G. Mason, B. Keys and S. Safe, Polychlorinated dibenzofurans (PCDFs): effects of structure on binding to the 2,3,7,8-TCDD cytosolic receptor protein, AHH induction and toxicity, Toxicol. 32:131 (1984).

86. S. Safe, Comparative toxicology and mechanism of action of polychlorinated dibenzo-p-dioxins and dibenzofurans, Ann. Rev. Pharmacol. Toxicol. 26:371 (1986).

87. H.K. Ouw, G.R. Simpson and D.S. Siyali, Use and health effects of Aroclor 1242, a polychlorinated biphenyl, in an electrical industry, Arch. Environ. Health 31:189 (1976).

88. A. Fischbein, M.S. Wolff, R. Lilis, J. Thornton and I.J. Selikoff, Clinical findings among PCB-exposed capacitor manufacturing workers, Ann. N.Y. Acad. Sci. 320:703 (1979).

89. M. Maroni, A. Columbi, G. Arbosti, S. Cantoni and V. Foa, Occupational exposure to polychlorinated biphenyls in electrical workers. II. Health Effects, Brit. J. Ind. Med. 38:55 (1981).

90. I. Hara, Health status and PCBs in blood of workers exposed to PCBs and of their children, Environ. Health Perspect. 59:85 (1985).

91. K.K. Steinberg, L.W.J. Freni-Titulaer, T.N. Rogers, V.W. Burse, P.W. Mueller, P.A. Stohr and D.T. Miller, Effects of polychlorinated biphenyls and lipemia on serum analytes, J. Toxicol. Envir. Health 19:369 (1980).

92. S. Safe, Polychlorinated biphenyls (PCBs) and polybrominated biphenyls (PBBs): biochemistry, toxicology and mechanism of action, CRC Crit. Rev. Toxicol. 13:319 (1984).

93. A.P. Alvares, A. Fischbein, K.E. Anderson and A. Kappas, Alterations in drug metabolism in workers exposed to polychlorinated biphenyls, Clin. Pharmacol. Ther. 22:140 (1977).

94. R. Warshaw, A. Fischbein, J. Thornton, A. Miller and I.J. Selikoff, Decrease in vital capacity in PCB-exposed workers in a capacitor manufacturing facility, N.Y. Ann. Acad. Sci. 320:277 (1979).

95. K.H. Chase, O. Wong, D. Thomas, B.W. Berney and R.K. Simon, Clinical and metabolic abnormalities associated with occupational exposure to polychlorinated biphenyls (PCBs), J. Occup. Med. 24:109 (1982).

96. A.B. Smith, J. Schloemer, L.K. Lowry, A.W. Smallwood, R.N. Ligo, S. Tanaka, W. Stringer, M. Jones, R. Hervin and C.J. Glueck, Metabolic and health consequences of occupational exposure to polychlorinated biphenyls (PCBs), Brit. J. Ind. Med. 39:361 (1982).

97. E.A. Emmett, Polychlorinated biphenyl exposure and effects in transformer repair workers, Environ. Health Perspect. 60:185 (1985).

98. R.W. Lawton, M.R. Ross, J. Feingold and J.F. Brown, Effects of PCB exposure on biochemical and hematological findings in capacitor workers, Environ. Health Perspect. 60:165 (1985).

99. D.P. Brown and M. Jones, Mortality and industrial hygiene study of workers exposed to polychlorinated biphenyls, Arch. Environ. Health 36:120 (1981).

100. A.K. Bahn, I. Rosenwalke, M. Herrmann, P. Grover, J. Stellman and K. O'Leary, Melanoma after exposure to PCBs, N. Engl. J. Med. 295:450 (1976).

101. K. Kreiss, M.M. Zack, R.D. Kimbrough, L.L. Needham, A.L. Smrek and B.T. Jones, Association of blood pressure and polychlorinated biphenyl levels, J. Am. Med. Assoc. 245:2505 (1981).

102. H.E.B. Humphrey, Evaluation of humans exposed to water-borne chemicals of the Great Lakes, Final report for EPA Cooperative Agreement CR807192 (1983).

103. K. Kreiss, Studies on populations exposed to polychlorinated biphenyls, Environ. Health Perspect. 60:193 (1985).

104. R. Bannister, D. Davis, T. Zacharewski, I. Tizard, and S. Safe, Aroclor 1254 as a 2,3,7,8-tetrachlorodibenzo-p-dioxin antagonist: effects on enzyme induction and immunotoxicity, Toxicol., in press.

105. J.M. Haake, S. Safe, K. Mayura, and T.D. Phillips, Aroclor 1254 as an antagonist of teratogenicity of 2,3,7,8-tetrachlorodibenzo-p-dioxin, Toxicol. Letters, in press.

106. D.L. Stalling, R.J. Norstrom, L.M. Smith, and M. Simon, Patterns of PCDD, PCDF and PCB contamination in Great Lakes fish and birds and their characterization by principal component analysis, Chemosphere 14:627 (1985).

107. J.J. Ryan, Variation of dioxins and furans in human tissue, Chemosphere 15:1585 (1986).

AN ESTIMATION OF THE ACTUAL HEALTH RISK OF FIREMEN AND WORKERS ASSIGNED
TO THE DECONTAMINATION FOLLOWING A FIRE INVOLVING PCB (ASKAREL AND CON-
TAMINATED OIL)

Gaétan Carrier

Hydro-Quebec - Direction Santé et Sécurité
75, west Dorchester blvd. - Floor 15
Montréal (Quebec)
H2Z 1A4

ABSTRACT

During a fire involving PCB, there is always a formation of extremely toxic products: the PCDF and PCDD. The literature concerning the
production of these toxic substances, our observations following the fire
at IREQ installations and epidemiological and clinical data are providing
us with some landmarks of the effects of these substances on human
beings. We try to estimate the actual risk for the firemen fighting the
flames, the workers assigned to the decontamination and for the population living in the surrounding of IREQ's installations and we indicate
efficient measures to prevent these risks.

INTRODUCTION

Many recent fires involving polychlorinated biphenyls (PCB) are
raising more and more concern in the general population, to the workers
and firemen. The most important problem is fires of electrical equipment
insulated with either "pure" PCB, or PCB mixed with other solvents like
the tri- and tetrachlorobenzenes, often present in the askarels. A related problem less serious but arousing a lot of worries is the electrical equipment insulated with mineral oils contaminated with PCB (from
concentration of a few ppm to more than 500 ppm). We estimate that
approximately 10 % of the transformers in service insulated with oil are
contaminated with more than 50 ppm.

The gravity of the fires involving PCB is due to the formation of
extremely toxic organochlorinated compounds under intense heat: the
polychlorodibenzofurans (PCDF) and the polychlorodibenzodioxins (PCDD).
We also find polychloroquaterphenyls (PCQ) and methylsulfones, their
toxicity is considered to being less important than the one of PCB and
the polychlorobiphenylenes; but there is very little documentation on the
toxicity of this last family.

When a fire occurs with equipment containing PCB - a transformer or
a condensator - the attention is put on security measures to take by the

firemen during the fire and by the workers assigned to the decontamination and repairing of the damaged area. It is important to know if these measures are efficient and this is what we will try to determine after assessing the risks involved.

Condition of formation for the PCDF, PCDD and hydrochloric acid

During the combustion of PCB, the potential of formation of PCDF and PCDD and their isomers depend on many factors: the products used (chlorine content, presence or lack of tetrachlorobenzenes), the temperature, retention time at an optimal formation temperature, oxygen content, catalysts presence in the other burned materials, etc.

The global behavior of PCB during thermal degradation is as represented in figure 1.

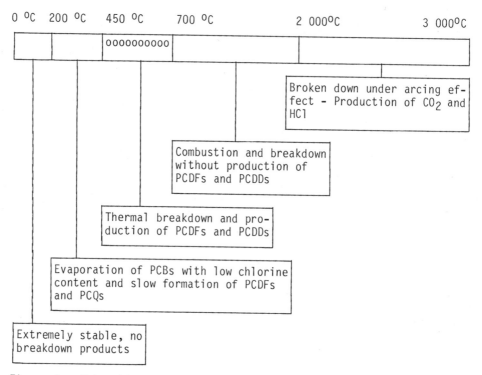

Figure 1. Schematic representation of the thermal breakdown of PCBs and the production of PCDFs, PCDDs and HCl

Let us analyse in greater detail the thermal breakdown of PCBs. From 0 to 200°C the PCB are extremely stable: there is no decomposition. From 200 to 450°C, even if these temperatures are smaller than the flashing point of the PCB, there is an evaporation of PCB and a slow formation of PCDF and PCQ. The formation of these products is function of the presence of oxygen and temperature [8], [16]. For example, Morita et al. (1978) have shown that for the AR 1248, in presence of oxygen, the optimal formation rate is reached at 300°C after 14 days (0,04 % of PCB converted in PCDF). At 270° and 330°C the quantities of PCDF were a hundred times less. The PCDF found in rice oil of Japan (1968) and Taiwan (1979) that provoked the Yusho and Yu-Cheng sickness was due, in both cases, to a contaminated heat exchanger with PCB of which the temperature varied from 240° to 360°C.

The flashing point of the PCB is around 450°C. During a fire, it is between 450° and 700°C that there is a formation of PCDF and PCDD. Many authors studied the thermal degradation phenomena at these temperatures for different types of PCB products, either pure PCB, PCB mixed with polychlorobenzene, contaminated oil or silicone with different concentrations of PCB [5], [10], [13], [16]. Under these conditions, the level of PCDF formed varies from a few milligrams per kg of PCB to a maximum of 40 g per kg of PCB. Erickson et al. [10] shown that with AR 1254, the optimum conditions are the following: $8\% < O_2 < 12$ %, $T° 675°C$, retention time at this temperature > 0.8 s. When one or many of these parameters are not near by these values, the quantity of PCDF diminish considerably. Moreover, in all these studies, it appears that isomers present are function of the type of PCB and there is a linear relationship between the number of PCB and the level of PCDF formed.

From **700 to 2000°C,** there is a production of PCDF in trace amount. Rappe et al. (1979)[18] shown that there is no detectable PCDF and PCDD when heating the PCB in a rotary ciment oven at temperatures between 1400 and 2000°C this indicating that this is a secure method of destruction for PCB.

At **2000°C and more,** and under the effect of electric arcs, there is a production of CO_2 and HCL: this phenomena can occur in a condensator when electric arcs are produced.

It is obvious that during a fire, we do not find necessarily all the optimum conditions of formation of the PCB found in the laboratory. The temperature varies in time and in space, with variations of many hundred of degrees Celsius between the periphery and the center of the fire. The molecules of PCB are not all subjected to the same combustion temperature. Moreover, it seems that certain molecules of PCB are lacking required to burn and form a molecule of PCDF. The PCB spread on the ground without being degraded as it is proved by the concentrations found in soot. We can deduct without too much risk of error that, during a fire, the quantities of PCDF formed will be much smaller than the maximum value found in optimum laboratory conditions.

Major fires of electrical equipment insulated with askarels

In North America, the worse fires involving PCB, and also the most notorious, are the one in Binghamton, in February 1981 in a government building of 18 floors, the one in San Francisco in May 1983 in a commercial building, and the one at IREQ, in November 1984. In these three cases, the soot covering the surface of the internal walls contained PCB, PCDF, PCDD and polychlorobiphenylenes. The most recent fire involving askarel occurred in Reims in France, January 14th 1985. [12]

Table 1 present the typical quantities of PCB, total PCDF, total TCDF and 2,3,7,8-TCDD coming from soot samples taken near by the seat of the fires in San Francisco, Binghamton and at the IREQ close by the burned electrical equipment:

In the three cases, the quantities of PCDF were relatively important; on the other hand in California and at IREQ, the quantities of total PCDD were negligible and the presence of 2,3,7,8-TCDD was not detected. The higher concentration found for PCDD and 2,3,7,8-TCDD at Binghamton can be explained by the presence of (35 %) of tetrachloro-

Table 1. Typical concentration of PCB, PCDF, TCDF and TCDD coming from soot samples taken near by the seat of the fire.

| FIRE | Concentrat. in ppm in the soot [mg/kg] | | | | |
	PCB	TOTAL PCDF	TOTAL TCDF	2,3,7,8-TCDF	2,3,7,8-TCDD
San Francisco (California) [15]	86 000**	286	16,29	6,70	not det.
Binghamton (New-York State) [15]	7 200 *	438	145,00	72,00	3
IREQ [21]	2 500**	39	4,89	0,61	not det. ***

```
*  :  AR 1254 and 35 % tri- and tetrachlorobenzenes
** :  100 % AR 1242
***:  Detection limit 10 to 50 mg/g
```

benzene in askarel. It is shown that with an incomplete pyrolysis tetra-chlorobenzene is decomposing in dioxins. The PCB have a greater tendency to produce furans. The mechanism of production of dioxins by pyrolysis of PCB is unknown.

It is worth noting that at IREQ and in San Francisco, the 2,3,7,8-TCDF represented respectively 1,5 and 2 % of the total PCDF, whereas at Binghamton it was approximately 16 %. This can probably be explained by the fact that the AR 1254 produce more isomers tetra-CDF than the AR 1242. At IREQ, more than 62 % of the PCDF were isomers tri-CDF. We then have to conclude that health risks are higher with AR 1254 than with AR 1242, and that the presence of tetrachlorobenzene in askarel increases the risk. Obviously, this depends on the burned quan-tities. Knowing more about the conditions of the fire at IREQ, we will make our estimations in relation to that specific case.

Estimation of the amount of PCDF formed during the fire at IREQ

Brief description of the fire at IREQ [21]. On November 20 1984, during a series of test performed on a power transformer in a section joining the high voltage laboratory (see figure 2), an explosion occurred in a bank of condensators. This explosion generated a fire that resulted in a series of explosion of other condensators, of which some were insu-lated with oil and other with PCB. A total of 2 560 L of mineral oil and 670 L of PCB (AR 1242) were destroyed. During the first two hours, of which most of all the oil and PCB was destroyed a dense black smoke stayed inside the building. It filled the annex, the centre hall and the spaces between the walls and the ceilings. After a while, a hole was burned through the ceiling on top of the seat of the fire where the smoke could escape. However, the samples taken outside near by the building, after the fire, show a negligible contamination of the soil. All the materials were fireproof, with the exception, of course, of the mineral oil and the PCB. A total of 3 230 L was destroyed (670 L of PCB and 2 560 L of mineral oil). Taking into account the densities of 0,9 for oil and 1,5 of AR 1242), we obtain a total of 3 309 kg of liquid.

A typical sample analysis of soot taken near by the seat of the fire indicates that it contains 39 ppm of PCDF (mg of PCDF per kg of soot) and 2 200 ppm of PCB (mg of PCB per kg of soot). The inside walls of the annex were covered with an oily layer of soot of a thickness estimated from a few micrometers to a ten millimeters, whereas the walls of the center hall were covered in certain spots of a thin film of soot of at most a few micrometers of thickness.

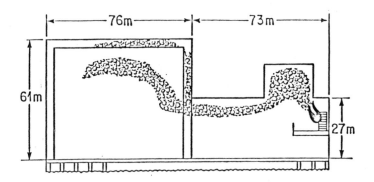

Figure 2.   Place of fire in the high voltage laboratory of IREQ (figure taken from David Train and Danielle Dupont's conference, given in Seattle (USA) in September 1985.

Quantity of PCDF formed. The walls, ceilings and floors of the annex and the hall are considered being smooth. Their total surface is approximately 45 000 m^2 (30 000 m^2 for the hall and 15 000 m^2 for the annex). According to a chemist from IREQ, Jacques Castonguay, who surveyed the decontamination, it is realistic to estimate that the thickness of the soot in the annex was of an average of a millimeter on all surfaces. He evaluates that taking into account the low density of the soot (0,010 to 0,015), that the weight of it is approximately 10 g/m^2 of surface. In the worst case, if the content of the soot per unit of surface of the hall was equal to the one of the annex, we would obtain a total of 450 kg of soot (45 000 m^2 x 10 g/m^2). This correspond to a total of 17,55 g of PCDF (450 kg of soot x 39 mg of PCDF/kg of soot). There was 670 L of PCB, hence a total approximately 1005 kg of PCB (density 1,5). We then obtain approximately 17 mg of PCDF per kg of PBC.

Let us verify by another way the quantity of PCDF that could be formed. We will use the hypothesis of the worst case. In ther first place, if we estimate that the concentration of soot found in all the building would be equal to the one measured near by the seat of the fire (which is an overestimation) and secondly, suppose that all the combustible (3 309 kg of combustible) would be transformed in an equivalent amount of soot (3 309 kg of soot). In this case, we would find a total of 129 g de PCDF (3 309 kg of soot x 39 mg of PCDF/kg of soot). Since, there has 1 005 kg of PCB, we obtain a maximum of 128 mg of PCDF per kg of PCB. It is then realistic to think that at IREQ the formation in PCDF per kg of PCB was smaller than this highly overestimated value.

Health risk during a fire involving askarel and during decontamination

Without pretending to estimate precisely the risks for acute and chronic toxicity of the exposed persons, we believe we can obtain some quite useful information by comparison with other events where the PCDF have been responsible for human intoxications. Since, in that area, there is two well documented cases of intoxication related to PCDF (in Japan in 1968 and Taiwan in 1979) [1], it seems to us that it would be quite interesting to seek to know if the risk is as important in the case of a fire involving PCB.

The Japanese named the sickness resulting from PCB intoxication the Yusho sickness, which means rice oil sickness and the Chinese called it Yu-Cheng sickness. This sickness was identified most exclusively with the chronic symptoms to the mucous membranes and skin chloracnee and also with alteration of the hepatic function. Although, in an acute phase, these persons were complaining of fatigue, loss of apetite, cephalea, weakness and oedemea to the extremeties. Nauses and vomitting were also mentioned but all those symptoms in the acute phase were considered to be minor complications [1], [14]. Notice that all the cases of intoxication with dioxin (2,3,7,8-TCDD) in Seveso, Italia (1976) have presented the same symptoms [17].

It is estimated that the victims of the Yusho and Yu-Cheng sickness have ingested an average of 0,5 to 2 g of PCB and 2,4 to 7 mg of PCDF [1],[6]. Moreover, we know today that the principal cause of the symptoms for these victims were the PCDF [1]; we will try hereafter to estimate if it is probable that a fire involving electrical equipment containing PCB can cause a similar risk as in the cases of Japan and Taiwan.

Risk during a fire. To estimate this risk, we are refering ourselves to the fire at IREQ. It has taken approximately six hours for the firemen to master the flames but it was in the first two hours that the fire reached its maximum intensity. We estimate that it is during that period that most of the PCB was burned. Just after the fire, a sample taken in the burned building, permitted to verify that the level of PCB in the air was smaller than $10 \mu g/m^3$. However, we do not have any data on the level of PCDF in the air at that moment.

Exposure to PCB. A human being breath approximately 2 m^3 of air per hour; for a fireman to inhale 0,5 to 2 g of PCB, the air would have to contain between 125 and 500 mg/m^3 of PCB. Taking into account the low vapor pressure of the PCB, even at temperature of 35 to45°C, it is almost impossible to reach such concentrations.

You should note that no volunteer fireman of Hydro-Québec nor any professionnal fireman had, after a week, levels of PCB in blood higher than the concentrations of the general population ($<$ 5 ppm of 1242 as an average). This is an excellent indication of the low risk at which they have been exposed. We have to say that a majority wore a positive pressure self-generated oxygen mask.

Exposure to polychlorodibenzofurans. From an estimation of the worst case (see section 3.2), a maximum of 129 g of PCDF could have been formed by the combustion of the PCB. The worse of the hypothesis is if the 129 g of PCDF formed gradually during the two hours and were spreaded uniformly in all the high voltage laboratory (an approximate value at 496 219 m^3: 76 x 76 x 61 + 73 x 73 x 27 meters), stayed in suspension in the air: in this case, the average concentration per m^3 would be equal to the average concentration after an hour, or approximately 0,125 mg/m^3 [½ x (129 g - 496 219 m^3 of air)]. After two hours, a fireman would have absorbed a maximum of 0,25 mg of PCDF, which is less than the absorbed quantity during the intoxication in Japan (2,4 to 7 mg). Obviously, this hypothetical case is unrealistic because as it is formed, the PCDF are trapped in the dense and oily smoke promoting their deposition on the ground and walls. The content of PCDF in the oily soot shows it; their concentration is much lower.

These observations permit us to conclude, that during a fire of a few hundreds of liters of PCB, the risk for those who fight the fire, to present symptoms of the Yusho sickness is negligible if the following safety measures are taken: all person needed to intervene in a fire involving equipment containing askarels should wear a positive pressure self-generated oxygen mask with overalls for protecting the skin. Indeed such a respirator represent a security factor of more than 3000. And even if the previous data indicated that the risk at Binghamton was at least ten times higher than at IREQ, the risk becomes negligible with such safety measures. Moreover we recommend to establish an emergency program for all areas where such equipment is found and in assuring that the access to the sinister be forbidden to all person not authorized during and after the fire.

It is important to mention that until now, after the fires involving askarel, there was no case of chloracnee reported, in both America and Europe.

Health risk during decontamination. This risk will be evaluated approximately by comparing with the quantities of PCDF that caused the Yusho sickness. Remember that the quantity of total PCDF coming from the samples taken near by the seat of the fires in San Francisco, Binghamton and IREQ were respectively 286, 438 and 39 mg per kg of soot. This implies that, to absorb a similar quantity of PCDF of the one of Japan (2,4 to 7 mg), the workers assigned to the decontamination would have had to absorb by inhalation, per contact or by ingestion a few grams of soot (see table 2). Even if the PCDF are not very volatile and that the soot is very oily and sticks to the surfaces, we have to say that the risk is quite high. It is indeed very important to take exceptional measures in order to prevent any contact with soot and any contamination of workers clothes [21]. The medical follow-up of our employees, during and after the decontamination, has shown that the measures taken at IREQ have been shown to be efficient [7]. Note that those safety measures were similar to the one taken during the decontamination in Binghamton.

Table 2. Quantities of soot to ingest for the absorption of total PCDF to be equivalent to the quantities causing the Yusho sickness.

FIRE	QUANTITY OF SOOT TO INGEST (G)
San Francisco	8,4 to 24
Binghamton	5,5 to 11
IREQ	60,0 to 175

As you can see, it is in Binghamton where the risk was the highest, according to two factors: the presence of tetrachlorobenzene and the high level of chlorine in AR 1254, comparatively to AR 1242 for the other two cases. In fact, the risk in Binghamton was even higher than it seems to be: indeed, the relative quantities of tetra-CDF and 2,3,7,8-TCDF (the most toxic isomers of the PCDF family) were higher than in the other cases.

Fire of equipment containing mineral oil contaminated with PCB

In the seventies, it has been found that the transformers insulated with oil and built since the beginning of the 50 were often contaminated with PCB. According to the inventory of Hydro-Québec, we estimate that 10 to 15 % of these transformers would have concentrations of more than 50 ppm of PCB, and less than a 1 % concentrations of more than 500 ppm. It is why the public, and particularly the workers in the electrical industry and the firemen, fear the consequence of a fire with such equipment. Is there a risk with the presence of PCDF during the combustion of these contaminated oils, and can we estimate it?

Quantities of PCB in the contaminated oils. With each part per million (ppm) of contamination with PCB correspond 1 mg of PCB per kilo of oil. Hence, a transformer which contains 300 kg of mineral oil contaminated with 100 ppm of PCB in weight, contains a total of 30 g of PCB (100 mg of PCB/kg of oil x 300 kg of oil), or a factor of 10^4 times less of PCB in comparison to a transformer insulated only with PCB.

Health risk for the PCDF formed during a fire of contaminated oil. Let us use again the example of the transformer containing 300 kg of contaminated oil at 100 ppm, or a total of 30 g of PCB. If we suppose that during the fire the PCB are destroyed in conditions similar to the fire at IREQ, we have the right to say that our estimation of the maximum amount formed of 128 mg of PCDF per kg of PCB is still valuable. We would then have a total of 3,8 mg of PCDF as a result of the combustion of the PCB present (100 molecules of PCB dispersed throughout 999 900 molecules of oil).

Imagine that these 300 kg of contaminated oil burn in approximately 30 minutes (quite realistic estimation); there would be an average emission of 2,0 μg/s of total PCDF over the oil in combustion at a certain time. If we take into account the fact that the fire will cause the particles to be sucked up in the air, the effect of dispersion of the wind and the effect of dilution with distance, it is easy to imagine that the concentration of PCDF per unit of volume will be quite low. Moreover, you have to take into account the fact that the PCDF are substances with low volatility: they have in fact a tendency to fall on the ground.

If we combine the average level of inhalation of a human being (2 m^3/h) to the duration of the fire (1 h), it is not very likely for a person to absorb significant quantities of PCDF. Due to safety measures (wearing of the positive pressure self-generated oxygen mask by the firemen and that the access be forbidden to all person not authorized), the potential health risk by the PCDF become pratically non existent in a case like that.

In October 1985, Eschenroeder Alan et al. [11] demonstrated with a laboratory study that burning 25 gallons (112,5 L) of oil contaminated with 9,9 ppm and with the use of a mathematical model of dispersion for the PCDF emitted, that the probability for a person in the surrounding of the fire to absorb an equivalent dose to the one indicated by the EPA for a life exposure (equivalent risk at 1/1 000 of the NOEL) was smaller than 1 in 7 000 000.

We think that these data, even if they are based upon approximate estimations, demonstrate that the health risk caused by a fire with contaminated oil is not at all of the same magnitude to a fire with askarels. However, the research should be pursue to assess precisely this risk. In the meantime, we recommend to be cautious: the firemen should wear the appropriate protection equipment during a fire. After the fire, even if the risk of contamination by the PCDF is low, it is necessary to act cautiously and to seek the advice of experts before starting cleaning operations and overhand of the area.

Conclusion

During a fire involving either pure or almost pure PCB, we estimate that, if safety measures are taken, like wearing positive pressure self-generated oxygen mask, gloves and overalls, the health risk for the firemen is negligible. With contaminated oil, even if the risk seems to be much smaller, we recommend the use of such measures. An emergency plan must also be implemented to move away any person not authorized from near by the dense smoke of the fire.

However the risk is higher for the workers in charge of the decontamination and the overhand of the area. It is why we insist that safety measures as the one taken at IREQ, which has shown to be efficient [7] be established. We must also make sure that the access to the sinister be forbidden to all person not authorized during and after the fire.

Before the cleaning of the area, an assessment must be done to evaluate the concentrations of the toxic products present and to determine the type of measures necessary for the decontamination in order to respect the long term exposure values for the products found [15], [21].

For our judgement, the replacement of the electrical equipment insulated with askarel is the best of the safety measures. It should be of the highest priority in the areas where the risk of contamination is high.

References

[1]   ***, "Japan, U.S. Joint Seminar: Toxicity of chlorinated biphenyls dibenzofurans dibenzodioxin and related compounds," Environmental Health Perspectives, 1985.

[2]   L. Abenhaim, "BPC et leurs produits de dégradation thermique: données contradictoires et problèmes décisionnels pour les industries électriques," Pollution atmosphérique, octobre-décembre 1985.

[3]   Ir.H. Aresu de Seui, "Les BPC et l'incendie," Association Nationale pour la Protection contre l'incendie Belgique, ISSN 0772-7267-February, 1985.

[4]   W. Boyd et al., "Managing PCB risks from transformer fires," 6 pages, April, 1985.

[5]   H.R. Buser et al., "Formation of polychlorinated dibenzofurans (PCDF) from the pyrolysis of PCBs," Chemosphere, pp. 109-119, 1978.

[6]   G. Carrier, "Les effets toxiques des BPC et des produits apparentés sur l'animal et l'humain," conférence présentée à la réunion de l'Association canadienne des médecins en santé au travail à Calgary, Canada, le 12 septembre 1985.

[7]   G. Carrier et D. Dupont, "Étude clinique d'une population exposée à des fumées lors d'un incendie impliquant des BPC et lors de la décontamination des lieux", conférence présentée à l'IEEE à Montréal, Canada, le 29 septembre 1986.

[8]   Chittim et al., "Chlorinated dibenzofurans anddibenzo-p-dioxins: detection and quantification in electrical equipment and their formation during the incineration of PCBs," Wellington Science Associates Inc. Prepared under contract n° OSS78-00067 for Fisheries and Environment Canada, Ottawa, 1979.

[9]   G. Eadon et al., "Comparisons of chemical and biological data on soot samples from the Binghamton State Office Building," Center for Laboratories and Research, N.Y. State Department of Health, 1982.

[10] M.D. Erickson et al., "Thermal degradation products from dielectric fluids," For environmental protection agency, Washington, D.C., (EPA), 1984.

[11] Alan Q. Eschenroeder and Edward J. Faeder, "Human health risks from PCB - contaminated mineral oil transformers," PCB seminar, Seattle, Washington, October 22-25, 1985.

[12] R. Fournié, "L'accident de Reims Premier bilan des analyses," Revue générale de l'électricité n° 1 - janvier 1986.

[13] O. Hutzinger et al., "Polychlorinated dibenzo-p-dioxins and dibenzofurans. A bioanalytical approach," Chemosphere, n° 10, pp. 19-25, 1981.

[14] T. Kashimoto et al., "Role of polychlorinated dibenzofuran in Yusho (PCB poisoning)," Arch. Environ. Health, vol. 36, n° 6, pp. 321-326, 26 refs, 1981.

[15] T.H. Milby et al., "PCB Containing Transformer fires decontamination guidelines based on health considerations," Journal of Occupational Medicine, vol. 27, n° 5, May, 1985.

[16] M. Morita, J. Nakagawa, C. Rappe, "Polychlorinated dibenzofuran (PCDF) formation from PCB mixture by heat and oxygen," Bul. Environ. Contam. Toxicol., n° 19, pp. 665-670, 1978.

[17] F. Pocchiari et al., "Human health effects from accidental release of TCDD at Seveso, Italy," Annals N.Y. Academy of Science, vol. 320, pp. 311-321, 1979.

[18] C. Rappe et al., "Dioxins, dibenzofurans and other polyhalogenated aromatics: production, use, formation and destruction," Annals New York Academy of Sciences, vol. 320, pp. 1-18, 1979.

[19] C. Rappe et al., "Occupational exposure to polychlorinated dioxins and dibenzofurans," Chlorinated dioxins and related compounds - Impact on the environment, O. Hutisinger, ed. Pergamon Press, Oxford, N.Y. "livre réservé", 1982.

[20] A. Schecter, "Contamination of an office building in Binghamton, New York by PCBs dioxins, furans and biphenylenes after an electrical panel and electrical transformer incident," Chemosphere, vol. 12, 4/5, pp. 669-680, 1983.

[21] D. Train, A. Chamberland, D. Dupont, J. Castonguay, "PCB capacitor fire at IREQ'S high voltage laboratory and subsequent decontamination," PCB seminar, Seattle, Washington, 1985.

ENVIRONMENTAL EFFECTS

BIOACCUMULATION OF POLYCHLORINATED BIPHENYLS IN CANADIAN WILDLIFE

Ross J. Norstrom

Environment Canada
Canadian Wildlife Service
National Wildlife Research Centre
Ottawa, Canada K1A OH3

INTRODUCTION

There are 209 possible structures arising from substitution of one to ten chlorine atoms on the biphenyl nucleus. Ballschmiter and Zell[1] have presented a shorthand form of numbering these components using rules developed by the International Union of Pure and Applied Chemistry (IUPAC), which has been adopted for this paper. This numbering system has gained a high degree of acceptance because it offers an alternative to arbitrary "Peak X" identifications based on order of elution on a chromatogram, and cumbersome structural formulae. The latter numbering system and stereochemically equivalent positions, ortho (o), meta (m) and para (p) are:

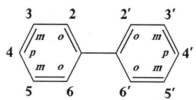

See Table 1 for cross-reference between structural formulae and IUPAC numbers of the major PCBs found in biological samples. This article follows the convention that a member of the PCB class of compounds is a congener, regardless of degree or position of chlorination; a sub-class of PCBs of specified degree of chlorination (e.g., tetrachloro) is a homolog, and a member of a homolog is an isomer.

Technical PCB products are usually identified by the percentage chlorine by weight in the mixture. The major North American manufacturer, Monsanto Corp., produced a series of PCB mixtures, the most common of which were called Aroclor 1242, 1248, 1254 and 1260, containing 42%, 48%, 54% and 60% chlorine by weight, respectively. These weight percentages approximately correspond to 3, 4, 5 and 6 chlorines per molecule, respectively. Aroclor 1254 and Clophen A50 (Bayer, W. Germany) contain mainly tetrachloro- to hexachlorobiphenyls, and Aroclor 1260 and Clophen A60 contain mainly penta- to heptachlorobiphenyls[2]. Some other major technical PCB mixtures are Kaneclors (Kanegafuchi, Japan), and Phenoclors (Prodelec, France); Sovol produced PCBs in the U.S.S.R.[2] Uses of PCBs have been voluntarily restricted to closed systems (e.g., capacitors) since the early 1970s in the U.S.A., but strong legislative restrictions did not come into effect until 1979[3]. PCB uses in Canada (where PCBs have never been manufactured), were regulated in 1977, and importation of PCB-containing equipment was banned after 1980 [4]. Historical uses of PCBs as additives in a number of products in combination with uncontrolled disposal of electrical equipment has resulted in massive contamination of the environment. Total world production to 1980 was in the order of 10^9 kg, of which about half was made by Monsanto Corp. in the U.S.A. [5]. The U.S. E.P.A. estimated that about 10^8 kg are in the environment[3]. Mackay[6] calculated that the distribution of PCB accessible for redistribution in the environment (atmosphere, soil, fresh water, biota) was about 10^7 kg, and a similar amount was deposited to marine sediments. Approximately one third of world production is still in use, and one third is in landfills or storage[6]. It has been calculated that 10^6 kg of PCBs cycle through the U.S.A. atmosphere annually[7]. The chemical stability which made PCBs so useful in the electrical industry has also made them a widespread environmental contaminant.

Although PCBs had been in use since 1929, it was only in mid 1960s that they were found to be major environmental contaminants[8]. By the early 1970s it had been established that PCBs were widely distributed in the environment, even in remote areas such as the Arctic[9], and in man[10]. Since that time a considerable body of research has accumulated on levels and dynamics of PCBs in the environment, and their toxicology [2,3 11-13]. One of the biggest hinderances to PCB research was the inability of earlier analytical methodology (i.e., packed column GC) to separate the components (congeners). This problem has been resolved by the advent of reproducible and easy to use fused-silica capillary GC columns. These columns are capable of such high resolution that two or three different liquid phases are sufficient to achieve separation of even highly complex mixtures. The limited number of phases which are popularly used facilitates comparison of chromatograms with those published in the literature.

Another analytical (and toxicological) problem was that only a few of the 209 PCB congeners required for confirmation of identity and calibration had been synthesized. Various alternative strategies were devised to identify individual PCBs. The most successful of these was a GC retention time index (RI) prediction based on the additivity of the indices for the two rings[14]. These half-RI values were obtained from symmetrically substituted PCBs and a few asymmetrical PCBs in which one ring had a substitution pattern with a known half-RI. There were still problems, however. One does not need to look hard in recent literature to come up with two or three alternative identities for some major GC peaks in PCB chromatograms.

Mullin et al.[15] recently reported on the synthesis and relative retention times of all 209 congeners on an SE-54 capillary GC column. Although there were some co-eluting compounds, the majority of the peaks were single component. With this information, identities can be established by matching with peaks obtained on other SE-54 and similar columns, such as the bonded phase DB-5 (J&W Scientific) column, which is probably the fused silica capillary column most employed in trace analysis of organochlorines. Certified quantitative standards of 52 PCBs, including most of the important components in technical mixtures, have also been introduced by the Marine Analytical Standards Program of the National Research Council of Canada, Halifax, N.S., and are available from them at a reasonable cost. With these advances, it is possible to make an assessment of individual PCB contaminantion in the environment. In fact, the biggest single problem now facing the PCB analyst is how to manage the huge data base that rapidly accumulates from the analyses.

This report gives an overview of the patterns of individual PCB congeners found in wildlife in Canada based on our latest research on birds at the top of aquatic food chains in the Great Lakes (Herring Gull), Pacific coast (Great Blue Heron) and Atlantic coast (Atlantic Puffin, Double-crested Cormorant, Leach's Storm Petrel and Gannet), and a terrestrial food chain (Peregrine Falcon). PCB pattern changes due to selective bioaccumulation in the Alewife - Herring Gull food chain in Lake Ontario and the fish - Ringed Seal - Polar Bear food chain in the Arctic are also presented. Trends in residue levels in Herring Gulls from the Great Lakes and Gannets and Petrels from the Atlantic coast are discussed.

METHODS

Eggs were used for determination of residue levels in birds except the Peregrine Falcon, for which only liver was avalable. In the case of Gannets and 1968/69 Petrels, archived samples from the Canadian Wildlife Service (CWS) National Specimen Bank were analysed. Details of the CWS organochlorine monitoring programs may be found in Mineau et al. [16], Pearce et al. [17] and Elliott et al. [18]. The large data base on organochlorine levels in seabirds accumulated by CWS since the late 1960s is summarized in Noble and Elliott[19]. Details on organochlorines in the Arctic marine ecosystem are given in Norstrom et al.[20,21] and Muir et al.[22]

Samples were analysed according to the procedure given in Peakall et al.[23] Lipids in Herring muscle, Ringed Seal blubber and Polar Bear fat were removed by Gel Permeation Chromatography[24] prior to separation into three fraction using Florisil. PCBs in birds and fish were determined using a 60 m DB-5 thin-phase column (J&W Sientific) and an electron capture detector (ECD) in a Hewlett-Packard 5840 GC equipped with an autosampler and splitless injection port. PCBs in Arctic samples were determined by GC/MS using a 30 m DB-5 column and a Hewlett-Packard 5987B instrument. Total ion chromatograms were obtained by monitoring the most intense ion in each of the M^+ clusters of tetra- to octachlorobiphenyls. The identity of individual congeners was established by matching order of elution of peaks in Aroclor mixtures from a DB-5 column to that determined by Mullin[25] using the same type of column, and from authentic standards (Marine Analytical Standards, National Research Council of Canada, Halifax, Nova Scotia).

PATTERNS OF PCB COMPONENT ACCUMULATION

Herring Gull and Alewife in the Great Lakes

The major PCB congeners found in Herring Gull eggs and Alewife, the principal source of organochlorine contamination in Herring Gulls, in Lake Ontario are listed in Table 1 along with bioaccumulation factors relative to the refractory PCB #153, which is nearly always the largest PCB component found in fish and higher organisms. The chromatograms of these two samples are compared to a mixture of Aroclor 1254:1260 (1:1) in Fig. 1.

Some substitution patterns, (2,6), (3,5), (2,4,6) and (3,4,5), are not commonly found in commercial mixtures. The number of chlorines on the two rings seldom differs by more than one for an odd number of chlorines in the molecule, and is usually equal for an even number of chlorines. A notable exception to this rule is #114, in which the number of chlorines on the two rings differs by 3. Because of these favoured

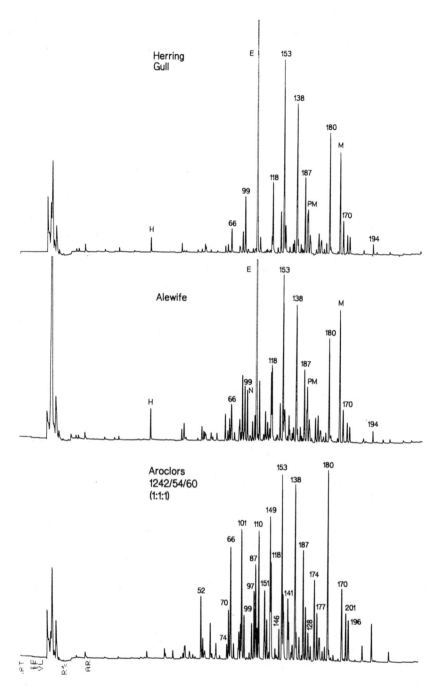

Fig. 1.   GC-ECD chromatograms (60 m DB-5 column) of the PCB fraction from a Herring Gull (egg), its main prey, the Alewife (a small forage fish) and a mixture of two technical PCBs (Aroclor 1254 and 1260). PCB congeners are identified by IUPAC No. (see Table 1), H is hexachlorobenzene, E is p,p'-DDE, N is t-nonachlor, PM is photomirex, and M is mirex.

Table 1. Bioaccumulation factor (wet weight basis) of PCB congeners from Alewife to Herring Gull eggs in Lake Ontario: dependence on chlorine substitution pattern.

IUPAC No.	Chlorine Substitution		Number of Unsubs. Adjacent Carbon		Biaccumulation Factor Relative to #153
	Ring 1	Ring 2	o,m	m,p	
52	2,5	2,5	0	2	<0.1
74	4	2,4,5	2	0	0.86
70	2,5	3,4	1	1	0.02
66*	2,4	3,4	2	0	0.43[a]
101	2,5	2,4,5	0	1	0.24
99*	2,4	2,4,5	1	0	0.74
97	2,3	2,4,5	1	1	<0.1
87	2,5	2,3,4	1	1	0.12
110	3,4	2,3,6	1	1	0.21
151	2,5	2,3,5,6	0	1	0.02
114	4	2,3,4,5	2	0	<0.1
149	2,3,6	2,4,5	0	1	0.18
118*	3,4	2,4,5	1	0	0.83
146	2,3,5	2,4,5	0	0	1.00
<u>153</u>	2,4,5	2,4,5	0	0	<u>1.00</u>
105*	3,4	2,3,4	2	0	0.57[a]
141	2,5	2,3,4,5	0	1	0.19
138*	2,3,4	2,4,5	0	0	0.93
178	2,3,5	2,3,5,6	0	0	0.52
187	2,4,5	2,3,5,6	0	0	0.94
128*	2,3,4	2,3,4	2	0	0.77
174	2,3,6	2,3,4,5	0	1	0.07
177	2,3,4	2,3,5,6	1	0	0.59
180	2,4,5	2,3,4,5	0	0	1.10
170*	2,3,4	2,3,4,5	1	0	1.05
201	2,3,5,6	2,3,4,5	0	0	0.97
196	2,3,4,5	2,3,4,6	0	0	1.07
194	2,3,4,5	2,3,4,5	0	0	0.90

* Indicates PCBs which induce "mixed" type oxidase enzyme activity in liver[12].

[a] Bioaccumulation factor is probably low due to presence of other congeners in the GC peak.

substitutions, the number of important congeners found in technical PCBs is far less than theoretical. Substitution at (2,5), (2,4), (2,4,5), (2,3,4), (2,3,4,5) and (2,3,5,6) with smaller contributions from (2,3), (3,4) and (2,3,6) accounts for most of the congeners in 50-60% chlorine technical PCB mixtures. In the less chlorinated mixtures, e.g., Aroclor 1242, di- and trichlorobiphenyl congeners with (2), (4), (2,4) and (2,5) substitution predominate[1].

Most fish lack enzyme systems necessary to metabolize PCBs with four or more chlorines per molecule into water-soluble, excretable metabolites. For example, Lieb et al.[26] detected no change in the Aroclor 1254 pattern in Rainbow Trout after 32 weeks on contaminated diet. In an an earlier study of PCB and other organochlorine contamination in Lake Ontario fish and gulls using a packed GC column and a Hall detector, we were able to simulate the PCB pattern found in small forage fish (Alewife and Smelt) and Coho Salmon quite accurately with mixtures of Aroclor 1254 and 1260[27]. There may be considerable species variability, however. Hinz and Matsumura[28], showed that *in vitro* metabolism of 2-2',5'-trichlorobiphenyl by Rainbow Trout and Channel Catfish liver microsomes was slow, but the rate in Goldfish was nearly equal to that in rats. A recent study indicated that Scorpion Fish from the North Sea had PCB patterns similar to that of birds and mammals, i.e., #74, #66 and #99 were the only early eluting PCBs not greatly diminished in comparison to seawater[29]. The pattern in Atlantic Cod was less altered than that in Scorpion Fish, but there was still a tendency for #74 and #99 to be higher. The overall PCB pattern in these cod was very similar to the one for Alewife in Fig. 1.

The main mechanism of PCB metabolism in birds and mammals is insertion of oxygen into two adjacent carbon atoms on the ring by hepatic microsomal oxidases (HMOs) to form a transient arene oxide intermediate, which in turn is degraded to hydroxylated PCBs[30]. The hydroxy-PCBs are conjugated to even more polar molecules and excreted in bile or urine. For rapid metabolism in most animals, including man[31], at least one adjacent pair of m,p carbon atoms must be unsubstituted, presumably because the large chlorine atoms hinder the approach of the enyzme. This rule certainly applies to Herring Gulls, since bioaccumulation factors for compounds with one or more unsubstituted pairs of m,p positions are 20% or less of that for #153 (Table 1). A secondary mechanism is oxygen insertion between the o,m positions. From Table 1 it appears that this mechanism is not particularly important in Herring Gulls, since #74, #99 and #118 bioaccumulated nearly as readily as #153. Rats have been shown to metabolize PCBs by direct insertion of a hydroxyl group, preferably at an unsubstituted m position, but this mechanism is usualy much slower[32]. An exception is the dog, which metabolizes #153 by this mechanism[33].

Increases in HMO enzyme activities can be induced by exposure to xenobiotics. In the case of organochlorines, there are two major classes that are •induced, the so-called phenobarbitol (PB) inducers and the methylcholanthrene (MC) inducers. The PB inducing PCBs (e.g., #153) have a low acute and chronic toxicity, whereas the MC inducers (e.g., 3,4,3',4'-tetrachlorobiphenyl) have toxicological properties similar to the highly toxic 2,3,7,8-TCDD[12]. Fortunately, the latter compounds are very minor in technical PCB mixtures. An intermediate class of PCBs which induce both enzyme systems, but with less potency, has been identified[12]. A number of these compounds are components in technical PCBs, including three of the four pentachloro PCBs with relatively high bioaccumulation factors, and the major component, #138 (Table 1). The *in vivo* toxicological significance of the so-called "mixed" inducers has yet to be established.

Fish-eating Marine Birds and Peregrine Falcons

Patterns of PCB congener accumulation in Great Blue Heron eggs from the Pacific coast, Gannet eggs from the Gulf of St. Lawrence and Peregrine Falcon liver from Ontario are indicated by the chromatograms in Fig. 2. Gannets and Herons are fish-eating, whereas the Peregrine Falcon only eats other birds, which it knocks from the air. Anatum Peregrines in eastern North America are endangered as a result of their sensitivity to thinning of egg shells by p,p'-DDE, the persistent metabolite of p,p'-DDT. Although Peregrines eat shorebirds if they are available, this specimen was probably preying mainly on terrestrial-feeding birds.

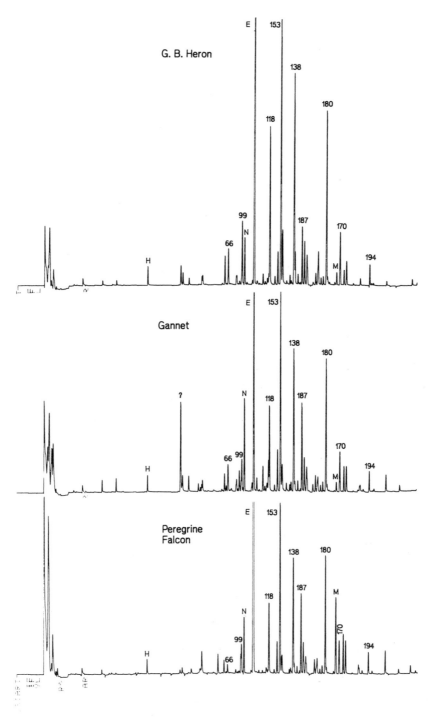

Fig. 2. GC-ECD chromatograms (60 m DB-5 column) of the PCB fraction from eggs of a Great Blue Heron (Pacific Coast), Gannet (Atlantic Coast) and liver of a Peregrine Falcon (Ontario). The first two species are fish-eating, the latter is bird-eating. Peak identities are as in Fig. 1.

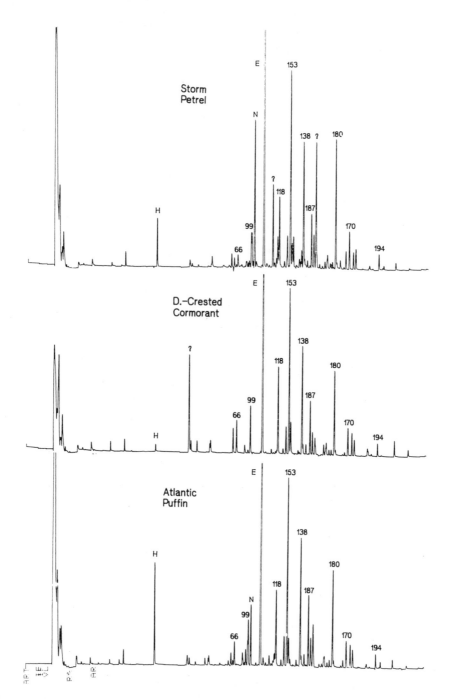

Fig. 3. GC-ECD chromatograms (60 m DB-5 Column) of the PCB fraction from eggs of three fish-eating birds from the Bay of Fundy on the Atlantic coast: Leach's Storm Petrel (offshore, surface feeding), Double-Crested Cormorant (nearshore, subsurface feeding) and Atlantic Puffin (offshore, subsurface feeding). Peak identities are as in Fig. 1. Peaks labelled (?) are unidentified.

The patterns are more remarkable for their similarity than their differences, considering that these specimens represent three species in three distinctly different ecosystems. The only tetrachlorobiphenyl isomers accumulated to any extent are #74 and #66, and pentachlorobiphenyl isomers are dominated by #99 and #118. Both the Gannet and Peregrine samples have unusually high levels of #206 (2,3,4,5-2',3',4',5',6'-nonachlorobiphenyl), the peak immediately following #194. This may indicate the presence of Aroclor 1268, 23% of which is #206 (M. Mullin, pers. commun.).

In order to show differences between species in the same general area, chromatograms for three species in the Bay of Fundy are presented in Fig. 3. The Double-crested Cormorant and Atlantic Puffin dive for fish, but the Puffin feeds more offshore. The PCB pattern in these species is similar to that in Great Blue Herons, Herring Gulls and Gannets. The Leach's Storm Petrel is unique in that it flies far out to sea and skims the surface for invertebrates and small fish. Levels in the Petrel are therefore expected to most closely represent atmospheric contamination. There are two peaks in the chromatogram (labelled with question marks) which have retention times similar to #151 and #128, respectively. From the relative abundance of these peaks, it is likely that these identities are incorrect. We have detected polychlorinated camphenes (probably dehydrochlorinated Toxaphene components) by GC/MS in the PCB fraction of fish from Baffin Bay. Because of the close connection between the diet of the Petrel and the atmosphere and the mainly invertebrate nature of the diet, exposure of the Petrel to toxaphene may indeed be higher than the other species, and it is likely that these peaks are Toxaphene-related.

Herring, Ringed Seals and Polar Bears in the Arctic

Because of interferences from Chlordane-related compounds in the PCB fraction [20-22], the PCB pattern in Arctic marine animals was determined using GC/MS. The chromatograms in Fig. 4 were generated using multiple ion monitoring, scanning one ion in each of the molecular ion clusters of tetra- to octachlorobiphenyls. This method also has the advantage of giving a more accurate representation of the relative amounts of PCB congeners, since there is less variation in sensitivity with degree of chlorination than with ECD.

The Herring chromatogram is remarkably similar to a mixture of Aroclors 1254 and 1260, although the ratio of the two is much higher than 1:1, as indicated by the relatively low level of #180. Of the major congeners, only #110 is significantly lower than expected. The pattern in the Ringed Seal is more like that in Figs. 1-3, and much the same rules of bioaccumulation therefore apply. The major differences between the patterns in seals and birds is the relatively higher level of (2,5) substituted components in seals, e.g., #52 and #101.

The PCB pattern in Polar bears is very unusual, and belongs in a category of it own. Over 95% of the total PCB is represented by 6 congeners: #99, #153, #138, #180, #170 and #194. Notably absent or much reduced from the pattern in seals are the normally refractory #74, #66, #118, #146, #105 and #187. The PCBs which accumulate in Polar Bears all have (2,4-2',4') minimum substitution. Most of the normally refractory PCBs which did not accumulate have an unsubstituted p position or more than 2 unsubstituted o positions. The Polar Bear therefore readily metabolizes PCBs with (4), (3,4), (2,3,5) and (2,3,5,6) chlorine substituents on one ring as well as those with unsubstituted m,p positions. Unlike the dog[33], the Polar Bear does not metabolize PCBs with unsubstituted m positions.

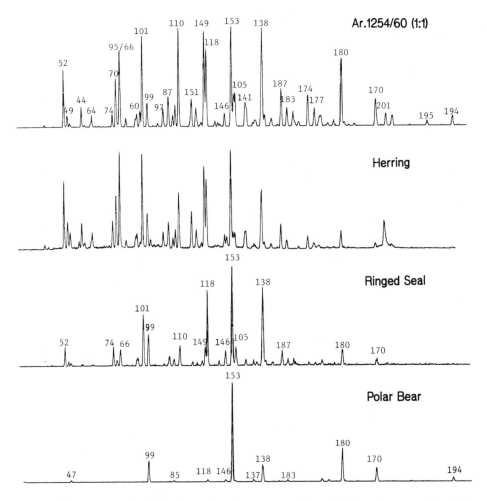

Fig. 4. GC-MS multiple-ion chromatograms (30 m DB-5 column, m/z 292, 326, 360, 394 and 430) of a technical PCB mixture (Aroclors 1254 and 1260), and of the PCB fraction from Herring, Ringed Seal blubber and Polar Bear fat from the Arctic. PCBs are identified by IUPAC No. (see Table 1).

LEVELS AND TEMPORAL TRENDS

Current PCB Levels

PCB levels in the individual specimens shown in Fig. 1, 2 and 3 are presented in Table 2. The levels are given as Aroclor equivalents, rather than the sum of individual PCBs, to facilitate comparison to earlier data. We have determined total organically bound chlorine in Herring Gull eggs from Lake Ontario by Neutron Activation Analysis, by total area in a packed column GC-Hall Conductivity Detector chromatogram and by packed column GC-ECD using two peaks, #153 and #138[34]. The results showed that PCB estimations by the latter method may be 37% too high. However, a more recent comparison of the sum of PCB congeners in

Table 2. Recent levels of PCBs (Aroclor 1254 + 1260 equivalents) and the apparent ratio of Aroclor 1254:1260 in Canadian wildlife.

Species	Type of Diet	Area	Year	PCB Level* (mg/kg)	Ratio, Aroclor 1254:60
Great Blue Heron	shallow water fish	Vancouver, B.C.	1983	3.57	2.7
Peregrine Falcon	birds (pigeons and smaller)	Ontario	1983	3.08	2.1
Herring Gull	omnivorous (~70% alewife)	Lake Ontario	1984	43.0	2.2
Alewife	aquatic invertebrates	Lake Ontario	1985	0.28	3.2
Gannet	pelagic marine fish	Gulf of St. Lawrence	1984	9.5	2.4
Leach's Petrel	open ocean surface biota	Kent Island, Bay of Fundy	1984	3.44	2.0
Double-Cr. Cormorant	pelagic marine fish	Manawagonish I. Bay of Fundy	1984	3.56	3.9
Atlantic Puffin	pelagic marine fish	Machias Seal I. Bay of Fundy	1984	3.20	3.0

* In eggs, fresh weight basis, except liver for peregrine falcon.

Herring Gull eggs to an Aroclor 1254:1260 equivalent calculated from a single peak (#138) in a capillary GC-ECD chromatogram indicated that there was little difference between the two numbers (CWS unpublished data). Further research is required to relate older data obtained by packed column GC to the sum of individual PCBs from capillary column GC.

PCB levels are similar in the Peregrine Falcon and all seabirds except the Gannet. Mean levels in Gannets are two to three times higher than those in seabirds from the Bay of Fundy and the coast of Newfoundland[17-19]. The level in Herring Gull eggs from Lake Ontario is significantly higher than in marine fish-eating birds, and has changed little since 1980[35].

The ratio of Aroclor 1254:1260 in Table 2 was calculated by the following formula:

$$\text{Aroclor } 1254{:}1260 = \frac{25.86R - 0.65}{7.95 - 1.7R}$$

where R is the ratio of Areas of #118 to #180 in the ECD chromatograms.

PCB #118 is a minor component in Aroclors 1242 and 1260, and is therefore diagnostic for Aroclor 1254, and #180 is diagnostic for Aroclor 1260[1]. The most useful diagnostic PCB for Aroclor 1242 is #28 (2,4,4'-trichlorobiphenyl), which was below the detection limit in all samples we analysed except for Herring in the Arctic, in which Aroclor 1242 peaks were relatively abundant. This is not indicated in Fig. 4 because trichlorobiphenyl ions were not included in the total ion chromatogram. The formula corrects for the minor contribution of Aroclor 1254 to #180 and Aroclor 1260 to #118.

In all samples the ratio of Aroclor 1254 to 1260 falls in the range from 2 to 4. It is interesting to note that ratio for the Double-Crested Cormorant is only 25% higher than for the Atlantic Puffin from the same area, although it is the highest ratio found among the birds in Table 2. Walker[36] found the *in vitro* hepatic microsomal oxidase activity in puffins to be an order of magnitude higher than in cormorants. The former species had activities similar to mammals, whereas the latter had activities similar to fish. The fact that we found neither substantially different PCB patterns nor significantly higher levels of PCB in Double-Crested Cormorants than in Atlantic Puffins indicates that very little if any metabolism of the more recalcitrant PCBs occurs in either bird. Excretion of unmetabolized PCBs probably occurs, however. The whole-body half life of the box-shaped perchlorohydrocarbon, mirex, which is not known to be metabolized by any species, is less than one year in adult Herring Gulls[37]. Excretion of unmetabolized organochlorines probably occurs via partitioning between the colon wall and fecal matter[38]. Depending on the number and weight of eggs a female lays, excretion in egg yolk may be important.

Temporal Trends

Temporal trends in PCB levels from 1968/69 to 1984 are presented in Fig. 5 for aquatic-feeding birds in four distinct aquatic environments in eastern Canada: Lake Superior, Lake Ontario, Gulf of St. Lawrence and the Atlantic Ocean east of Newfoundland. In order to make a better comparison, the levels are given relative to peak level in the time period, or the earliest available data. Peak levels were 75 mg/kg and 250 mg/kg in Herring Gull eggs from Lake Superior (1974) and Lake Ontario (1972), respectively, 24 mg/kg in Gannet eggs from the Gulf of St. Lawrence (1969) and 3.4 mg/kg in Leach's Storm Petrel eggs from the North Atlantic (1972).

Lake Superior is a large, deep, oligotrophic lake with a long water residence time. PCB input is mainly from the atmosphere[39]. The watershed of Lake Ontario is highly industrial and urban, in contrast to Lake Superior, and ranks as one of the most organochlorine-contaminated bodies of water in the world, largely due to effluent from chemical industries in the Niagara Falls, N.Y. area[40], which accounts for PCB levels 3 times higher than in Lake Superior. Gannets are feeding in a marine environment, but close to shore, whereas the Petrels are feeding far from shore. Thus, there is a distinct gradient in PCB contamination with lacustrine > nearshore marine > offshore marine environments. The inshore Gulf of St. Lawrence environment may be influenced by residues coming from the Great Lakes and further downstream in addition to long-range atmospheric transport. PCBs in the Atlantic Ocean are due to long-range atmospheric and ocean current transport.

In spite of the large differences in environments and history of contamination, declines occured in all four cases between 1972 and 1980, then levelled off. During the 1972 to 1980 period even the *rates* of decline were similar, in the order of 15% per year. The voluntary restrictions on PCB use and decreases in emissions to the environment in the early 1970s therefore resulted in relatively rapid and significant reductions in PCB contamination in both highly contaminated and remote environments. However, let us not forget that current levels of PCB in

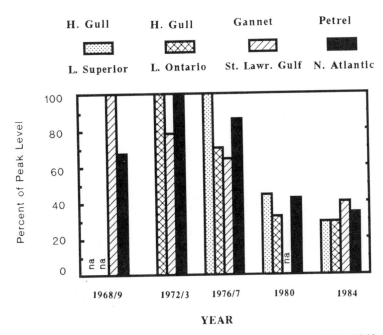

Fig. 5. Long-term trends of PCB levels (Aroclor 1254:1260, 1:1, equivalents), in eggs of aquatic-feeding birds from the Great Lakes (Herring Gull), the Gulf of St. Lawrence (Gannet) and the North Atlantic Ocean (Leach's Storm Petrel). Data are presented as percent of peak level; na = not analysed. Peak level for Herring Gull in Lake Superior is based on earliest data available, 1974.

Lake Ontario Herring Gulls are still 10 times higher than peak levels in Leach's Petrels, and show no signs of decreasing. Global-scale recycling of PCBs between the atmosphere and land and ocean surfaces is probably reaching an equilibrium, and even without further input to the environment, we cannot expect the rapid improvements of the 1970s to continue. PCBs will be with us for many years to come, although we will probably see gradual shifts in individual components to more highly chlorinated isomers, such as we have observed in Gannet eggs [18].

ACKNOWLEDGEMENTS

I am grateful to Henry Won, Mary Simon and Derek Muir for providing the chromatograms, and John Elliott, Peter Pearce and Chip Weseloh for data from the seabird and Herring Gull monitoring programs. I also wish to acknowledge the many biologists and technical staff who were involved in collecting and preparing samples for analysis.

REFERENCES

1. K. Ballschmiter and M. Zell, Analysis of polychlorinated biphenyls (PCB) by glass capillary chromatography, *Fres. Z. Anal. Chem.* 302:20 (1980).

2. O. Hutzinger, S. Safe and V. Zitko, "The Chemistry of PCB's," CRC Press, Cleveland (1974).

3. National Academy of Sciences, "Polychlorinated Biphenyls," : National Academy of Sciences, Washington, D.C. (1979).

4. D.L. Grant, Regulation of PCBs in Canada, Ch. 27, *in*: "PCBs, Human and Environmental Hazards," Butterworth Publishers, Boston (1983).

5. R.L. Durfee, G. Contos, F.C. Whitmore, J.D. Barden, E.E. Hackman and R.A. Westin, "PCBs in the United States - industrial use and environmental distributions," EPA 560/6-76-005 (NTIS No. PB-252-012), U.S. E.P.A., Washington, D.C. (1976).

6. M.D. Erickson, Physical, chemical, commercial, environmental and biological properties, Ch. 2 *in*: "Analytical Chemistry of PCBs," Butterworth Publishers, Boston (1986).

7. T.J. Murphy, L.J. Formanski, B. Brownawell and J.A. Meyer, Polychlorinated biphenyl emissions to the atmosphere in the Great Lakes region. Municipal landfills and incinerators, *Environ. Sci. Technol.* 19:942 (1985).

8. S. Jensen, A.G. Johnels, M. Olsson and G. Otterlind, DDT and PCB in marine animals from Swedish waters, *Nature* 224:247 (1969).

9. A.V. Holden, Monitoring organochlorine contamination of the marine environment by the analysis of residues in seals, *Marine Pollution and Sea Life,* Fishing News (Books) Ltd., London (1970).

10. S. Jensen and G. Sundstrom, Structures and levels of most chlorobiphenyls in human adipose tissue, *Ambio* 3:70 (1974).

11. F.M. D'Itri and M.A. Kamrin, eds., "PCBs, Human and Environmental Hazards," Butterworth Publishers, Boston (1983).

12. S. Safe, Polychlorinated biphenyls (PCBs) and polybrominated biphenyls (PBBs): biochemistry, toxicology, and mechanism of action, *in*: "CRC Critical Reviews in Toxicology," CRC Press, Cleveland (1985).

13. NRCC, "Polychlorinated Biphenyls: Biological Criteria for an Assessment of Their Effects on Environmental Quality," NRCC No. 16077 National Research Council of Canada, Ottawa, Ont. (1978).

14. D. Sissons and D. Welti, Structural identification of polychlorinated biphenyls in commerical mixtures by gas-liquid chromatography, nuclear magnetic resonance and mass spectrometry, *J. Chromatog.* 60:15 (1971).

15. M. D. Mullin, C.M. Pochini, S. McCrindle, M. Romkes, S.H. Safe and L.M. Safe, High-resolution PCB analysis: synthesis and chromatographic properties of all 209 PCB congeners, *Environ. Sci. Technol.* 18:468 (1984).

16. P. Mineau, G.A. Fox, R.J. Norstrom, D.V. Weseloh, D.J. Hallett and J.A. Ellenton, Using the Herring Gull to monitor levels and effects of organocholorine contamination in the Great Lakes, Ch. 19 in: "Toxic Contaminants in the Great Lakes," John Wiley and Sons, New York (1984).

17. P.A. Pearce, J.E. Elliott, D.B. Peakall and R.J. Norstrom, "Organochlorine contaminants in eggs of seabirds in the northwest Atlantic, 1968-1984," Canadian Wildlife Service manuscript (in prep.).

18. J.E. Elliott, R.J. Norstrom and J.A. Keith, "Organochlorines and eggshell thinning in Gannets from eastern Canada, 1968-1984," Canadian Wildlife Service manuscript (in prep.).

19. D.G. Noble and J.E. Elliott, "Environmental contaminants in Canadian Seabirds, 1968-1985," Technical Report Series No. 13, Canadian Wildlife Service, Ottawa, Canada (1986).

20. R.J. Norstrom and D.C.G. Muir, Long-range transport of organochlorines in the Arctic and Sub-Arctic: evidence from analysis of marine mammals and fish, in: "Proc. World Conf. on Large Lakes, Mackinac, Mich., May 19-21," (in press, 1987).

21. R.J. Norstrom, M. Simon, D.C.G. Muir and R.E. Schweinsburg, Organochlorine contaminants in Arctic marine food chains: identification, geographical distribution and temporal trends in Polar Bears, (submitted to Environ. Sci. Technol., 1987).

22. D.C.G. Muir, R.J. Norstrom and M. Simon, Organochlorines in Arctic marine food chains: accumulation of specific PCB congeners and chlordane-related compounds, (submitted to Environ. Sci. Technol., 1987).

23. D.B. Peakall, R.J. Norstrom, A.D. Rahimtula and R.D. Butler, Characterization of mixed-function oxidase systems of the nestling Herring Gull and its implications for bioeffects monitoring, Environ. Toxicol. Chem. 5:379 (1986).

24. R.J. Norstrom, M. Simon and M.J. Mulvihill, A gel-permeation/column chromatography method for the determination of CDDs in animal tissue, Intern. J. Environ. Anal. Chem. 23:267 (1986).

25. M. Mullin, Workshop on High Resolution PCB Analysis, Large Lakes Research Station, U.S. E.P.A., Grosse Ile, Michigan, May 1985.

26. A.J. Lieb, D.D. Bills and R.O. Sinnhuber, Accumulation of dietary polychlorinated biphenyls (Aroclor 1254) by Rainbow Trout (Salmo gairdneri), J. Agr. Food Chem. 22:638 (1974).

27. R.J. Norstrom, D.J. Hallett and R.A. Sonstegard, Coho salmon (Oncorhynchus kisutch) and Herring Gulls (Larus argentatus) as indicators of organochlorine contamination in Lake Ontario, J. Fish. Res. Board Can. 35:1401 (1978).

28. R. Hinz and F. Matsumura, Comparative metabolism of PCB isomers by three species of fish and the rat, Bull. Environ. Contam. Toxicol. 18:631 (1977).

29. V. Weigelt, Kapillargaschromatographische PCB-musteranalyse mariner Spezies - Betrachtungen zwischen Anreicherung und chlor Substitution ausgewahlter PCB-componenten in marinen Organismen aus der deutschen Bucht, Chemosphere 15:289 (1986).

30. G. Sundstrom, O. Hutzinger and S. Safe, The metabolism of chlorobiphenyls - a review, Chemosphere 5:267 (1976).

31. F.M.S. Wolff, J. Thornton, A. Fishbein, R. Lilis and I.J. Selikoff, Disposition of PCB congeners in occupationally exposed persons, Toxicol. Appl. Pharmacol. 62:294 (1982).

32. S. Kato, J.D. McKinney and H.B. Matthews, Metabolism of symmetrical hexachlorobiphenyl isomers in the rat, Toxicol. Appl. Pharmacol. 53:389 (1980).

33. I.G. Sipes, M.L. Slocumb, D.F. Perry and D.E. Carter, 2,4,5,2',4',5'-hexachlorobiphenyl: distribution, metabolism and excretion in the dog and monkey, Toxicol. Appl. Pharmacol. 65:264 (1982).

34. R.J. Norstrom, A.P. Gilman and D.J. Hallet, Total organically-bound chlorine and bromine in Lake Ontario Herring Gull eggs, 1977, by instrumental neutron activation and chromatographic methods, Sci. Total Environ. 20:217 (1981).

35. R.J. Norstrom, T.P. Clark and D.V. Weseloh, Great Lakes monitoring using Herring Gulls, Ch. 8 in: "Hazardous Contaminants in Ontario: Human and Environmental Effects," Institute for Environmental Studies, U. of Toronto, Toronto, Ontario (1985).

36. C.H. Walker, Species variations in some hepatic microsomal enzymes, Progr. Drug Metab. 5:113 (1980).

37. T.P. Clark, R.J. Norstrom, G.A. Fox and H.T. Won, Dynamics of organochlorines in Herring Gulls (Larus argentatus): II. A two compartment model and data for ten compounds, Environ. Toxicol. Chem. (in press, 1987).

38. T. Rozman, E. Scheufler and K. Rozman, Effect of partial jejunectomy and colectomy on the disposition of hexachlorobenzene in rats treated or not treated with hexadecane, Toxicol. Appl. Pharmacol. 78:421 (1985).

39. S.J. Eisenreich, B.B. Looney and J.D. Thornton, Airborne organic contaminants in the Great Lakes ecosystem, Environ. Sci. Technol. 15:30 (1981).

40. J.O. Nriagu, ed., "Toxic Contaminants in the Great Lakes," John Wiley and Sons, New York (1984).

DEVELOPMENT OF DECONTAMINATION GUIDELINES FOR PCB/PCDF AND PCDD DECONTAMINATION IN AREAS OF HIGH EXPOSURE POTENTIAL

Richard L. Wade
Executive Vice President

Med-Tox Associates, Inc.
1431 Warner Ave., Suite A
Tustin, California 92680

Increased recognition of the spread of Polychlorinated Biphenyls (PCB's), PCDF's (Polychlorinated dibenzofurans), and Polychlorinated dibenzo Dioxins (PCDD's) following transformer, capacitator fires and failures have resulted in industry and government attention to decontamination efforts. Two critical questions which surface after most spill events which in turn result in PCB/PCDF/PCDD spread to adjoining surfaces are:

(1) What is the degree of exposure of individuals incidentally exposed to these products; and what risk do they face as a result of this exposure?

(2) What level of decontamination would be necessary to allow the facility to return to unrestricted use?

Individuals involved in these events typically include:

(1) Government officials
(2) Responsible parties (owners, buildings managers, etc.)
(3) Clean up contractors
(4) Consultants

For the above individuals, the first questions they face following a PCB/PCDD/PCDF spill include what is the risk to individual's exposure at the spill site and what level of residual PCB's/PCDD's/PCDF's are acceptable.

The most common form of major PCB releases are related to either transformer or capacitor failures. These PCB and related products can result from either the transformer or capacitator being externally heated by a fire, internal arking from internal shorting, which may result in a

pressure release or a fire within the transformer. The
rapid increase in temperature of the PCB's within the
transformer vessel can result in a mist of PCB's and
related products being sprayed for great distances. PCB's
an their dilutants, such as benzene, can also be combusted,
crating soot and acid fumes being widely disseminated.

"How Clean is clean?"

In establishing safe levels "How clean is clan", several
approaches have been used by regulatory agencies. The
first approach is to set a "safe" level based on informed
judgment, expert opinion, expert advisory committees, etc.
This approach of consensus standards has been used for
years by regulatory agencies, national testing
laboratories, professional associations and responsible
industries.

This approach often has the disadvantage of being a
qualitative decision, based on available data and
circumscribed by the experience of the participants.
Written rationales are often missing, on less than
adequate, for further peer or legal review. The consensus
approach does, however, allow for a strong introduction of
"rational" common sense, qualitative risk assessment, etc.
to be introduced into the decision making process. Many of
our existing health and safety environmental standards were
developed on this approach.

A second approach is to set environmental decontamination
standards at levels that are no greater than compatible
non-contaminated levels (background) Using this approach,
one seeks a standard based on not what is safe but depends
on what is non-contaminated. This approach does not take
into account questions of cost, benefits, relative health
or environment risks or any problems associated with the
decontamination effort. A number of questions related to
sampling and analytic sensitivity and reliability do,
however, surface and are critical to reaching "safe" levels
in once contaminated environments. In ascertaining
background levels for PCDD's and PCDF's it is often
difficult to obtain permission to sample areas for
background determination.

The third approach is to set standards based on a "non-
detectable" criteria. This approach then becomes totally
dependent on sampling and analytic sensitivity and
reliability. Health or environmental impacts, nor cost vs.
benefits issues are of little concern to the decision
making process. This approach does, however, assure
maintenance of a pristine environment.

Within the last five years, regulatory agencies and other
parties have started using more quantitative methods to
determine the relative risks associated with residual
levels of contamination found after an environmental spill
or release such as a PCB transformer or capacitator
failure.

The basic scientific approach for utilization of quantitative toxicological risk assessment techniques can be used to help address questions of risk associated with varying exposure assumptions.

The same scientific approach can be used for addressing two questions

    (1)   How clean is clean   –   A risk assessment to evaluate potential risk given exposure variables

    (2)   What is the risk to persons exposed on a site   –   An exposure assessment

For both risk assessments and retroactive exposure assessments, the use of available information to quantify the area of exposure "potential" is critical. Likewise, the model selected and its hypothesis used to quantify risk as a function of exposure can significantly affect the outcome of any exposure risk assessment. Therefore, care should be given to model selection, variable identification and assumption enumeration.

The general formulation for these assessments is that Risk is a function of Exposure x Potency.

Thus, one must strive to ascertain the exposure opportunity for individuals that may be exposed (a risk assessment) or retroactive where exposed (an exposure assessment).

Likewise, one has to agree upon a model which defines the potency of a given delivered dose of that chemical to produce a specified adverse biological response or effect.

Given the range of assumptions that can go into both the exposure and potency components of the equation, it is not unusual that significant differences can exist between two parties conducting a risk assessment on the same chemical situation or event. In order to help minimize these variables the USEPA has issued generic guidelines on the use of and format for both risk and exposure assessments (1,2,3).

For the purpose of discussion, this paper will breakdown some of the critical assumptions made in the derivation of acceptable risk as it relates to (PCB-PCDF potency and exposure). EPA has proposed use of its formula:

    P = g* X
    g = potency
    X = exposure – to reflect the risk faced by individuals exposed to dioxin and Furan contaminants found in PCB's

X has been determined to range from 2.0 pg/kg/day to 0.006 pg/kg/day. (See Table I)

TABLE I
TOTAL DOSE ACCEPTABLE RISK LEVELS
USED IN RISK AND EXPOSURE ASSESSMENTS

USEPA	.06 pg/kg/day
State of California Health Department	2 pg/kg
State of California Air Resource Department	.06 pg/kg/day
New York State	2 pg/kg
Canada	0.2 - 1.0 pg/kg/day
Netherlands	4 pg/kg/day
CDC	.028 - 1.40 ng/kg/day

EPA uses the 0.006 pg/kg/day 2,3,7,8 - TCDD equivalents although several states have accepted 2.0 pg/kg/day. European countries us acceptable daily intake levels ranging from 1.0 pg/kg/day to 10 pg/kg/day.

In establishing clean up levels or "how clean is clean", the traditional starting point is to determine an acceptance risk end point.

There is also no agreement by regulatory agencies on the acceptance of risk. EPA has accepted 1×10^5 risk in several risk and exposure assessments. Whereas some regulatory agencies consider any risk greater than 2×10^{-6} as being socially unacceptable.

Once the responsible authority in charge of the clean up has been established and an acceptable risk level established the calculations of exposure potential, model selection and risk calculations can commence. Thus, one selects a desirable outcome and back calculates to develop the acceptable levels of residual PCB, PCDF/PCDD than can be allowed which would not generate excess and unacceptable risks.

At this stage, a model must be selected which will reflect the dose/response relationship. The range of models available project widely ranging dose/response curves with some being linear whereas others are non-linear. It is beyond the scope of this text to provide detail as to the proper model selection. The USEPA guidelines do, however, provide criteria for model selection.

Once a model has been selected, consideration must be given towards the relative toxicity information on the various PCB/PCDD/PCDF isomers.

Most scientists agree that 2,3,7,8- TCDD is most potent as evidenced by animal carcinogenity testing, cell keratinization, and an enzyme inhibition biossays. As the ring structure becomes increasingly saturated beyond the four chlorine ring structure, the acute toxicity appears to decrease. Likewise, the cl atom in the 2,3,7,8 positions appears to be critical determinant in acute toxicities (4), (5). The exact mechanisms of acute 2,3,7,8 toxicity are still largely unknown. The dibenzofurans are considered to be significantly less toxic than the dioxin, with potency being (0.2-0.33) x 2,3,7,8 - TCDD.

The only two isomers of TCDD's that have undergone complete biossay are 2,3,7,8 TCDD and two congeners of hexachlorinated dibenzo droxins 1,2,3,6,7 and 1,2,3,7,8,9 hexachlorinated debenzo dioxins. These two hexa isomers were subsequently found to be contaminated with 0.03% TCDD.

Within the U.S. the USEPA and centers for disease control (CDC) and others have relied on traditional approaches to equate annual biossay data to human exposure situations. These methods extrapolate risk from animal tumor data from high dose exposures to low dose human exposures. The model most commonly used is the linearized multistage model, which assumes the agent in question is a tumor initiator. Regulatory authorities in some European countries and Canada do not rely upon tumor initiator models, but believe TCDD to be a tumor promotion as opposed to an initiator[6]. Such factors as these can change the assessment risk involved in low dos exposures by several orders of magnitudes.

Thus, model selection, and data selection to define dose/response curves, are critical variables in assessing risk. The guidelines on carcinogenic risk assessment issued by EPA[7] in 1984 give considerable freedom to the risk assessor on model and data selection and results on the same problem assessed by two individuals can vary widely.

Just as the model selection can vary widely, so can the assumptions that go into assessing exposures. Factors such as environmental half life of the agent in question (PCB, PCBF, PCDD's inside buildings); the opportunity for exposure to contaminated surfaces, inhalation rates, airborne concentrations based on actual air measurements or mathematical estimation, etc. all can significantly affect the exposure side of the equation and thus affect risks.

Inhalation Rates: In evaluating the opportunity for exposure, one critical factor is the inhalation rate. The value of air inhaled can vary from <10 M^3 for sedentary workers with an eight hour exposure opportunity in excess of 24 M^3 for 24 hour exposure opportunities. The inhalation rate selected is important and can be a significant variable for assessing total delivered dose of this sid. Most assessments assume 100% efficiency of lung absorption of inhaled PCB/PCDD, PCDF contaminants.

Dermal Exposure Opportunity: In assessing dose and risk for PCB's and PCDD's/PCDF's, a critical variable is the opportunity for dermal exposure. In determining dermal exposure opportunity it is customary to differentiate between areas within a given space that are readily assessable for high level contact such as exterior vertical and horizontal surfaces vs. areas less directly accessible. In these areas the person with potential exposure was high probability of dermal contact. This is in contrast to areas inside mechanical systems or in areas that because of height or inaccessibility allow for less opportunity for dermal exposure.

Typically data as to levels of PCB's/PCDD's/PCDF's are reported in units per 100 cm_2 or M^2. These data can then be separated into area of high exposure opportunity and loss exposure opportunity.

Another variable is the area of area of skin that can come in direct skin contact with these surfaces. Some of the original work on this issue was done by the New York State Department of Health. Under the New York State method, the activity is evaluated as to exposure potential and the surface area in cm^2 specified for individuals and tasks[6]. Assessments are then made as to the percent of uptake off surfaces. Empirical evidence suggest that 80-90% of the surface PCB contamination can be removed by wiping a surface with a hexane saturated wiping cloth. It is thus seen as a very conservative assessment that the dry skin of the hand would have significantly less uptake efficiency[7]. Variations on this uptake efficiency can significantly vary the outcome.

Another variable is the percent of dermal penetration by the PCB's/PCDD/PCDF. Here the efficiency of penetration has ranged from 10-100%. There is little empirical data available, and many regulatory agencies have accepted 108 dermal penetration efficiency of PCB through the skin.

Ingestion: Most of the risk and exposure assessments done on PCB's resulting from transformer/capacitator contaminat- ions have not considered ingestion as a significant source of exposure. In risk and exposure assessments related to soil contamination and in residential settings, ingestion does become a significant factor. The U.S. Center for Disease Control has established guidelines on exposure assessments for TCDD contaminated soils[10].

Based on the previously mentioned model variables and assumptions Table 1 shows the wide range of acceptable risk levels of dose per kilogram per day. doses at these levels are considered by these respective regulatory agencies to not result in risk greater than those seen as acceptable by that body, traditionally 1 x 10-5 or 1.0 x 10^{-6}.

TABLE II

TABLE OF BACKGROUND LEVELS OF
POLYCHLORINATED BIPHENYLS

AIR SAMPLES	UG/M^3
OUTSIDE TRANSFORMER VAULTS[1]	.001 − .09
OUTSIDE TRANSFORMER VAULTS[1]	.02
INSIDE SCHOOL VAULTS[1]	.002 − .004
INSIDE SCHOOL VAULTS[1]	.001 − .004
AFTER BALLAST BURNOUT[1]	14 − 166
LABORATORY[2]	.2 − .24
OFFICE[2]	.08 − .10
OUTSIDE AIR (CITY)[1][2]	.0018
HOME KITCHENS	.18 − .58
HOME LIVING ROOMS	.039
OUTDOORS	.04
BASEMENT	.012
GARAGE	.06
BEDROOM	.17

WIPE SAMPLES	UG/100 CM^2	MEAN	.SD
OFFICE BUILDINGS			
CINCINNATI	<.05 − .45	.08	.09
ST. PAUL	<.01 − .48	.13	.12
BOSTON	<.01 − .32	.07	.07
TOTAL	.01 − .48	.09	.10

SAN FRANCISCO

OFFICE BUILDINGS[3] <.01 − .20

---

[1] JOHN KOMINSKY, JAMES MELIUS, JEROME FLESCH: PRESENTATION AIHA CONFERENCE, PHILADELPHIA, PA, 1983. "EXCESSIVE CONTAMINATION FROM ELECTRICAL EQUIPMENT FAILURES"

[2] KATHRYN MACLEOD, "PCB'S IN INDOOR AIR. ANALYTIC CHEMISTRY BRANCH USEPA, USEPA DOCUMENT 560/6-75. RESEARCH TRIANGLE PARK, NC 27711

[3] DATA COLLECTED AS PART OF THE ONE MARKET PLAZA TRANSFORMER FIRE, SAN FRANCISCO, 1983

TABLE III

REPRESENTATIVE REPORTED ENVIRONMENTAL
LEVELS OF TOTAL PCDD'S/PCDF'S

LEVELS		NOTES
(1) AIR		
INSIDE	ND - 10,000+ PG/M^3	INSIDE BUILDING 30 DAYS FOLLOWING PCB TRANSFORMER FAILURE
OUTSIDE	.09 - 1.3 FG/M^3	(IN AREA WITH NO EVIDENCE OF CONTAMINATION)
WATERSTREAM	1 NG/KG - 3 UG/KG -	AFTER 3 SILVEX SPRAYING (335 DAYS)
INCINERATOR EMISSIONS	ND - 440 NG/M^3	
(2) WATER		
GROUNDWATER SURFACE RUNOFF	MOST STUDIES REPORT ND 0.05-0.06 PPT	(IN AREAS 42 DAYS AFTER SILVEX APPLICATION1 [6]
(3) SOIL (ND (<.1 PPB) - 1800+) PPB		
(4) SURFACES	<.01 NG/M^2 - > 2000 NG/M^2	
(5) DUST (SEE INCINERATOR ABOVE)		(IN SOME BUILDINGS DUST, AND IN OTHERS VAPOR PHASE CONTAMINATION PREDOMINATE)
(6) PCB OILS	PCB OILS (ND - 40 PPM) (TOTAL PCDD/PCDF) (IN NON-FAILED TRANSFORMERS)	

TABLE III (continued)

REPRESENTATIVE REPORTED ENVIRONMENTAL
LEVELS OF TOTAL PCDD'S/PCDF'S

LEVELS		NOTES
(7) PLANTS	ND – 40 PPB	MOST STUDIES SHOW NO PLANT UPTAKE FROM SOILS CONTAMINATED WITH LOW LEVELS OF 2,3,7,8-TCDD (POSSIBLE SEVESO EXCEPTION)
(8) SOLID WASTE	ND – 440 NG/M^2 (TOTALS EXTRACTABLE PCDD/PCDF) 4	

ENVIRONMENTAL LEVELS

In contrast to the use of risk assessment techniques to determine decontamination levels of PCB's, PCDD/PCDF's, some regulatory bodies have either said clean to background levels or taken background levels when setting allowable exposures. Tables (II and III) give some levels of background PCB's and PCDD/PCDF's.

Table III gives some levels of TCDD – the wide range indicates that some hazard levels of TCDD/PCDD and PLDF exists in most biological tissues with exposure species and individuals showing elevated levels.

Environmental Decomposition Factors: Photodegradation of 2,3,7,8- TCDD in areas of intense sunlight has been demonstrated to be significant (4). However, most risk assessments done to date place a half life from 5-10 years on PCB/PCDD/PCDF's found inside buildings or in areas on nondirect sunlight.

Tables IV and V show mix regulatory agency acceptable levels of PCDD/PCDF in soil, fish, air, water and surfaces.

Table VI lists a series of PCB's PCDD/PCDF decontamination standards that have been established for decontamination of selected sites. Most of the standards established in Table V were established as a result of risk assessments.

TABLE IV

TISSUE LEVELS PCDD/PCDF's

BIOLOGICAL	LEVELS	NOTES
CATTLE	4-40 PPT FAT	CATTLE GRAZING ON CONTAMINATED SOILS
BOVINE MILK	<40 PPT TO 7.9 PPB	SEVESO: NORMAL SILVEX APPLICATION RATES DID NOT RESULT IN ANY SIGNIFICANT INCREASE IN MILK 2,3,7,8-TCDD LEVELS
HUMAN MILK	ND (<1) TO 805 PPT	
FISH	ND - 695 PPT	(SIGNIFICANT VARIATION DEPENDING ON LOCATION CAUGHT)
HUMAN ADIPOSE	5-10 PPT TETRA DIOXINS 17 PPT PENTA FURANS 17 PPT HEXA 13 PPT HEPTA 600-800 PPT OCTA DIOXIN	

SOIL
(SOIL (SURFACE TCDD) <20-1929 PPT) - SUM TCDD <10-5237 PPT
SOIL (SUBSURFACE TCDF) 1413-23,920 PPT) - SUM TCDF ND-13,809 PPT

TABLE V

GUIDELINES USED IN VARIOUS ENVIRONMENTAL EXPOSURE SETTINGS FOR PCDD/PCDF

FISH
NEW YORK HEALTH DEPARTMENT
ONTARIO
FDA

10 PPT
20 PPT
25 PPT

SOIL

1 PPB - CDC

AIR

10 PG/M^3 NYS INDOORS (8 HOURS EXP)
10 PG/M^3 CALIFORNIA INDOORS (8 HOURS EXP)
3 - 28 NG/M^3 - INSIDE BUILDINGS
10 PG/M^3 CANADA

WATER
MICHIGAN/OUTFALL TO RIVER

10 PARTS PR QUADRILLION

TABLE VI

DECONTAMINATION EFFORT	STANDARD ACCEPTED FOR DECONTAMINATION		DOSE BASIS OF DEVELOPMENT RISK ASSESSMENT
	SURFACE	AIR	
BINGHAMTON			
PCDD/PCDF	3.3-28 NG/M^2	10 PG/M^3	2 PG/KG
PCB'S	1.0 UG/100 CM^2	1.0 UG/M^3	
SAN FRANCISCO			
PCDD/PCDF	3 NG/M^2 (ABOVE BACKGROUND)	10 PG/M^3	2 PG/KG
PCB	1.0 UG/100 CM^{1}3		
COLUMBUS, OH			
PCDD/PCDF	20 NG/M^2*	10 PG/M^3	2 PG/KG
PCB	1.0 UG/ 100 CM^2	1.0 UG/M^3	
SANTA FE			
PCDD/PCDF	1.0 NG/M^2	.06 PG/KG	2.0 PG/KG
PCB	0.5 UG/10 CM^2		
TULSA, OK			
PCDD/PCDF	3.3-38 NG/M^2	10 PG/M^3	2 PG/KG
PCB	1.0 UG/ 100 CM^2	1.0 UG.M^3	
CANADA			
PROD/PCDF	25* - 100 MG^2	10 PG/M^3	0.2-1.0 PG/KG/DAY (M O D E L DEPENDENT)
PCB	1.25 UG/ 100 CM^2		

* HIGH CONTACT SURFACES

CONCLUSIONS

Development of site specific risk and exposure assessments can be expensive but the cost is decreasing as these techniques become more routine. The commonly made assumptions for such assessments are becoming more standardized, yet considerable variations still occur in the potency factors applied to individual cogeners of PCDD's/PCDF's. Likewise, significant variations exists as to the potency of 2,3,7,8- TCDD (0.06 - 2.0 pg/kg/day) as acceptable intake levels.

There have been few recent advances in the toxicological literature as to the hazards of 2,3,7,8- TCDD or its equivalents. At a recent European conference reported human pharmocokinetic studies indicate the biological half-life of 2,3,7,8- TCDD has been calculated to be on the order of five years. Such conclusions were interpreted as justifying the rethinking of the reduction of acceptable daily intake levels. TCDF data on the biological fate of PCDF's support the five year biological half-life theory.

Increased reliance is being made of the existing risk assessment documents and the resulting standards that have been developed. Thus, future risk assessment for PCB's, PCDD's, and PCDF's likely be on hold until further toxicological evidence is developed. Advances in such risk assessment methods will likely be done as a result of the critiques of the existing risk assessment. The same principles of risk assessment and exposure assessment are being routinely used to evaluate individual's risk following incidental exposure.

Site specific exposure assessments, however, are powerful tools to be used to modify existing guidelines or document quantatively evaluate risk of persons either occupationally or incidentally exposed to PCB's, PCDF's, PCDD's, etc.

REFERENCES

1   Federal Register Friday, November 23 1984, Proposed Guideline for Carcinogen Risk Assessment.

2   Federal Register Wednesday, January 9, 1985, Proposed Guideline for Health Risk Assessments of Chemical Mixtures.

3   Endangerment Assessments for Superfund Enforcement Actions, Mogan et al. Support Branch Office of Waste Programs Enforcement, U.S. EAP 1984. Address request for information to WH527, 401 M Street, SW, Washington, D.C. 20460.

4   R. J. Kochiha et al - Comparative Toxicity and Biological Activity of Chlorinated Dibenzo-P-Dioxins and Chlorinated Dibenzo Furans Relative to 2,3,7,8- TCDD, Chemosphere J.Vol. 14, pp. 649-60, 1985

5   Eve Roberts – <u>The Ah Receptor and Dioxins Toxicity from Rodent to Human Tissues,</u> Chemosphere Vol. 14, No. 6/7, pp. 661-674, 1985

6   H.P. Shu, D.J. Paustenbach and F.J. Murray "A Critical Evaluation of the Use of Mutagenosis, Carcinogenesis and Tumor Production Data in Cancer Risk Assessment of 2,3,7,8 TCDD <u>Regulatory Toxicology & Pharmacology,</u> <u>Vol. 7, 1987</u>

7   EPA (1984b) Proposed Guidelines for Carcinogenic Risk Assessment (Chem. Regulation Rep. PP. 1025-1056)

8   Nancy Kim, Health Risk Assessment of PCB exposure New York Department of Public Health, Publication, March, 1983.

9   Sampling rewipe from PCB decontamination project IT Comp San Francisco, California, 1984.

10   Kimbrough, R.; Falk, H.; Stehr, P. and Fries, G., Health Implications of 2, 3, 7, 8 TCDD contamination of soil, J. Toxicology Environmental Health 14, 47-93.

DECONTAMINATION
AND RETROFILLING

STATE-OF-THE-ART TECHNOLOGY

FOR PCB DECONTAMINATION OF CONCRETE

John P. Woodyard

Director, PCB Management Services
International Technology Corporation
Torrance, California

Enzo M. Zoratto

Project Manager, Chemical Decontamination
International Technology Corporation
Pittsburgh, Pennsylvania

INTRODUCTION

Decontamination of building structures and equipment containing polychlorinated biphenyls is rapidly becoming a major concern to environmental managers in industry and government. The most costly stumbling block to this type of decontamination and the focus of more technical attention than any other aspect of building decontamination is the decontamination of concrete.

The technology required today in polychlorinated biphenyl (PCB) decontamination situations, including decontamination of concrete, was not readily anticipated when the original PCB regulations were passed in 1978. This need has spawned the development or application of numerous techniques for decontamination of surfaces and materials containing PCB. Whether driven by emergency situations, concerns over liability or complicated employee health risk issues, PCB decontamination in many ways serves as the testing ground for chemical decontamination technology to be used in the future for other chemical contaminants.

However, most of the technology currently in use has been developed for other applications and brought to PCB decontamination. Because they are not always applicable to active work environments or other more sensitive locations such as building interiors, these technologies have been adapted to fit the need posed by PCB spill and fire incidents in buildings.

This paper describes the development and application of technologies for removing PCB from concrete. The paper begins with an overview of the characteristics of concrete and the types of contamination that may be experienced. Forces driving the decontamination of concrete are reviewed including regulations, health risk concerns and environmental liability. The various techniques currently in use are then surveyed including wet and dry washing techniques, concrete removal and in situ treatment. The selection and application of these techniques are then discussed with a focus on factors such as cost, waste generation, production rate and cross-contamination potential.

CONCRETE CONTAMINATION

Concrete contamination with PCB can occur in several ways and take many forms. The characteristics of the concrete combined with the nature of the contamination dictate the selection of decontamination techniques and the extent to which they must be used.

Characteristics of Concrete

Structural and ornamental concrete has several attributes which combine to make this substance unique from a decontamination standpoint. First, all concrete is porous to some degree, thereby allowing penetration by either liquid or vapor contaminants. Second, concrete continuously ages and dries over its life span, changing its porosity and its ability to absorb contaminants. Third, and perhaps most important, concrete is often an integral structural component of the building, so decontamination must be performed; demolition and disposal is not always an option.

Types of Concrete Contamination

Four basic types of PCB concrete contamination are encountered in practice. They are:

• Surface soot
• Penetrated fumes
• Surface liquid
• Penetrated liquid.

Surface soot is most usually associated with PCB fires. The particulate associated with an electrical equipment or other type of fire involving PCB carries the PCB to various surfaces in the building. The soot remains on the surface until either the PCB is vaporized, the soot is blown or otherwise removed from the surface, or soot particles are washed or ground into the concrete.

Penetrated fumes are also generated during the course of many PCB fires and are actually the predominant mode of PCB transport in instances where concrete is dry and more porous. These fumes will penetrate below the outward surface of the concrete and condense in the interstitial voids, making the PCB and other combustion products more difficult to remove.

Surface liquid is probably the most common form of PCB contamination, since spills are the most common PCB-related incident. Surface liquid contamination can result from spills that are immediately cleaned up or from tracking of PCB by firemen or others involved in the response. PCB transported or placed in such a manner rarely penetrate beyond 0.25 inches and more often adhere to the concrete on or just below the immediate surface.

Penetrated liquid is the most troublesome form of PCB contamination. When allowed to stand on horizontal surfaces, PCB will readily penetrate concrete. PCB penetration is commonly encountered in buildings where PCB was routinely used in an unenclosed fashion for electrical equipment manufacture or when spilled from otherwise closed systems and not completely cleaned up. Penetration of PCB to over 12 inches has been documented in the absence of fissures or other means of transport. In such cases, the typical resolution is concrete removal and disposal.

FORCES GOVERNING CONCRETE DECONTAMINATION

Once PCB contamination is in place, its removal can be governed by several forces:

- Emergency spill or fire cleanup
- Concern for long-term environmental damage
- Concern for short- or long-term human health
- Short- and long-term liability from human exposure.

Emergency spill or fire cleanup involves the cleanup of PCB which was legally used in an enclosed manner but suddenly became open to the environment. Small PCB spill and fire cleanups are almost considered routine in businesses that still own and operate PCB equipment, and proper techniques for spill response and cleanup are well documented.

Concern for long-term environmental damage from PCB is most often associated with former disposal and spill sites rather than concrete contamination. Since PCB contained in or on concrete is generally considered even less mobile than when contained in soils or water, PCB in concrete is not considered as great an environmental risk. PCB decontamination of concrete is therefore rarely driven by broad-based environmental concerns.

Indoor decontamination is more commonly driven by concern for short- or long-term human health than for environmental protection. This concern for human health is often prompted by the discovery of former PCB spills and the presence of incidental contamination in older facilities. Human health exposure is usually associated with surface or airborne contamination, neither of which is typically associated with concrete contamination. Nonetheless, routine tracking or spillage of PCB on concrete will result in the need for concrete cleaning or removal in older plants where employees could still be exposed.

Short- and long-term liability from human exposure to PCB contamination of concrete is a manifestation of a perceived health risk. The mere presence of PCB in the work environment or a building's structural members is considered a long-term financial risk to the ultimate owner. State laws such as the Environmental Cleanup Responsibility Act (ECRA) in New Jersey have listed PCB as a contaminant of concern and have forced owners selling property to assure the state and buyer that contaminants such as PCB do not remain. Initial investigations revealing the presence of PCB will often precipitate both an ECRA cleanup and employee concerns over exposure. Property owners will then effect rapid and thorough decontamination, rather than risk the liability associated with employee exposure.

PCB DECONTAMINATION STANDARDS FOR CONCRETE

PCB decontamination standards for concrete are usually a function of the concerns mentioned above. Table 1 lists some of the more commonly cited guidelines for PCB decontamination. No regulations currently address this issue with specific standards.

In situations where health risk concerns are of greatest importance, surface and air contamination standards will often predominate. This is particularly true in commercial office buildings where PCB contamination is present in paint or concrete on walls, furniture and equipment. These are considered high contact areas for employees and, along with air, represent the most direct route of human exposure. Because of the porous nature of concrete, standard solvent-based wipe sampling techniques are not as applicable to concrete surfaces as they would be to bare metal surfaces and other nonporous materials.

Environmental cleanup standards for PCB are not applicable to concrete because the pathway/receptor assumptions made in the development of environmental standards do not apply as readily as they do to soil/ground water systems. PCB cleanup standards expressed in parts per million (ppm) generally assume some mobility of the PCB and dilution in the ultimate receiving water before reaching animals or humans. For interior applications in particular, dilution and mobility are not considered significant issues. Such standards are nonetheless applied when other standards cannot be agreed upon (Woodyard and Wade, 1986).

For states enacting ECRA-type legislation, the principal concern to the governing agency involves PCB left in place rather than either its mobility in the environment or human exposure. Standards used in this situation are typically a combination of health and environmental standards involving both ppm and $ug/100\ cm^2$ values. Decontamination standards have therefore been applied in some instances to both penetrated material and surfaces. Under New Jersey ECRA, for example, the working PCB standards are expressed in ppm for 0.5 inch increments. Decisions involving PCB contamination within concrete are often negotiable but are rarely based on scientifically developed risk assessment.

DECONTAMINATION TECHNIQUES

Given a particular type of concrete contamination and an associated driving force for decontamination, one or more methods must be chosen to meet the standard. The selection of a technique is based on several factors including production rate, effectiveness, potential for cross contamination, potential for enhanced penetration, waste disposal or treatment requirements and cost.

Decontamination techniques currently available in the industry can be categorized under one of the following major applications:

• Dry removal of soot
• Wet removal/washing
• Surface concrete removal
• Deep concrete removal
• In situ treatment.

In situations where the depth and extent of contamination are well known, appropriate technique selection is often evident. Under many circumstances, however, the depth and extent of contamination is not completely known at the onset of the project; the proper technique is therefore selected as the project proceeds and the extent and depth of contamination is better defined.

Dry Removal of Soot

Undisturbed soot particles and contaminated dust and dirt adhering to the surface of concrete are relatively easy to remove. The most frequently applied techniques for soot removal include vacuuming and the application of strippable coatings.

High efficiency particulate absolute (HEPA) vacuums are usually used for soot and dust removal. These vacuums are fitted with both submicron particulate filters and carbon canisters to collect the PCB soot particles and vapor respectively. Care must be taken to ensure that PCB vapor generated by the vacuuming process is not re-entrained in the air where it is more difficult to remove and presents a greater health hazard.

Strippable coatings have been used in the nuclear industry to remove radioactive particles from concrete surfaces and have found similar application in PCB decontamination. The coating, a special latex or comparable paint mix, is applied by hand in a thick layer. After drying, the coating, along with the underlying soot particles, is removed by hand. The use of coatings is most appropriate only when the soot is completely undisturbed. Therefore, coatings offer less flexibility than HEPA vacuuming (International Technology Corporation, 1986).

Wet Removal/Washing

Surface washing techniques traditionally have been among the most common methods used to decontaminate concrete surfaces. Most such techniques are rooted in the industrial services industry where they have been used for decades for tank and ship hull cleaning. Wet techniques are typically characterized as water-based, detergent/ water or solvent-based mixtures.

Water-based techniques consist of hydroblasting or pressure washing using only water. These methods rely entirely on inertial impaction to remove the solid or liquid PCB from the surface, as PCB has a low solubility in water. The water is applied to the surface under controlled conditions and subsequently collected for treatment, typically using carbon adsorption.

Detergent and water mixtures can be applied in the same situations as water, using either high pressure techniques or direct surface wiping with gauze. The most commonly applied detergents include Penetone, Citrikleen and Triton-X. Each of these detergents has a 1 to 5 percent solubility for PCB and therefore provides some measure of PCB chemical removal. However, because high pressure washing relies on inertial impaction, the presence of detergent provides only the ability to retain PCB in suspension for ease of collection. Use of pressure washing with detergent should also be considered carefully in view of the significant water treatment requirements and the difficulty and complexity of treating some mixtures. Detergents, therefore, are more prudently applied using wipe techniques.

Solvent washing of concrete surfaces is a wipe technique performed almost exclusively by hand using gauze or other means. Solvents have the advantage of providing increased PCB solubility but are more expensive and difficult to control and recover. Furthermore, because of the solubility enhancement, indiscriminant use of solvents for PCB washing can drive PCB further into the concrete or increase the airborne concentration of PCB through co-vaporization. Paint stripping solvents have been successful in removing PCB-contaminated paint from concrete surfaces by hand, without the housekeeping problems associated with using dusty, dry paint removal techniques.

Surface Concrete Removal

The type of PCB contamination of concrete that is most frequently encountered and most difficult to treat is just below the surface, resulting either from short-term spill contact or grinding of dry deposits through tracking. PCB has been shown, however, to penetrate concrete to much greater depths than common sense would indicate, through capillary action or other physical mechanisms. Decontamination of floors in particular sometimes shows PCB penetration of 0.5 to 2.0 inches for even small spills of limited contact time.

Dry surface removal techniques can be characterized according to the depth of concrete they are capable of removing. In order of increasing depth of removal, techniques most commonly used for this purpose include:

- Sandblasting
- Shot blasting
- Grinding or scarifying
- Scabbling
- High pressure water cutting.

These and other similar techniques have been adapted primarily from road maintenance and nuclear decontamination, where most of the prior experience exists.

Sandblasting is primarily a surface treatment that allows removal of small quantities of concrete from the surface but is more applicable to removal of stains or surface deposits. Compared to most methods, sandblasting offers little control -- even with vacuum collection devices -- due to the high speed of application and the unpredictable size grading and deflection pattern. Sandblasting can result in significant redeposition of PCB on other surfaces through sand or in air.

Shot blasting is a more refined version of sandblasting and is usually applied to floor surface treatment. Shot blasting leaves a relatively smooth surface suitable for immediate surface treatment. The depth of removal varies as a function of shot blaster speed and shot size. Most shot blasters are fitted with vacuum collection devices. These allow the recovery of the shot and concrete dust,

TABLE 1

SOME COMMONLY CITED STANDARDS FOR PCB DECONTAMINATION

LIMIT	AGENCY	DATE	EXPOSURE SCENARIO	APPLICABILITY & RESTRICTION
<100 ug/100 cm^2	Consensus	05/17/85	Restricted access areas.	For spills of fluid of 10 pounds or more PCB weight.
<100 ug/100 cm^2	Consensus	05/17/85	Non-restricted access areas.	For spills of one pound or more of PCB by weight; utility poles and asphaltic materials consider constituent PCBs.
10 ug/100 cm^2	Consensus	05/17/85	High contact ares; residential yards, sidewalks, playgrounds, and roadways.	Spills of PCBs onto solid surfaces, asphalt, metal, concrete, wood poles.
100 ug/100 cm^2	Consensus	08/23/85	Reduced public access areas; industrial facilities, rural roadways, rural fields and oil fields, also, restricted access electrical substations.	Spills of PCB materials onto solid surfaces or electrical equipment and machinery prior to redisribution in commerce.
0.5 ug/100 cm^2	NIOSH	11/10/83	Presumably inferior surfaces.	Based on upper background limit in non-manufacturing facilities.
1 ug/100 cm^2	Dept. of Public Health, City and County of San Francisco	11/17/83	General occupancy. Safe re-entry levels for decontaminated areas, office building.	Agency health-based cleanup criteria determined by risk assessment for One Market Plaza.
10 ug/100 cm^2			Restricted access. Transformer vault (assuming 2 hours per month occupancy).	

(Continued)

TABLE 1
(Continued)

LIMIT	AGENCY	DATE	EXPOSURE SCENARIO	APPLICABILITY & RESTRICTION
1 ug/100 cm²	New York Dept. of Health	1981	Office building, general exposure. PCB value used as indicator for presence of TCDD/TCDF.	Agency health-based cleanup criteria determined by risk assessment.
1.0 mg/m³ @ 42% Cl PCB 0.5 mg/m³ @ 54% Cl PCB	OSHA	Jan. 1981	PEL - 8 hg (Permissible exposure level)	The higher the percent chlorinated of the PCB the lower the PEL allowed. PEL designed to provide protection from systematic intoxication Agency standard does not view PCBs as carcinogens.
1.0 mg/m³ @ 42% Cl 0.5 mg/m³ @ 54% Cl	ACGIH	1985 - 1986	TLV-TWA - 8 hrs (Time weighted average)	No respiratory protection required. TLV-TWA is based on acute or non-carcinogenic effects. Protective clothing and eye protection required.
1.0 ug/m³	NIOSH	Sept. 1985	TLV - 10 hrs	TLV is based on PCBs regulated as occupational carcinogens. At or below this limit no respiratory protection required.
>50 ppm	EPA	1983	Disposal at or above this concentration is regulated. Disposal of PCBs <50 ppm is unregulated and currently pending further rulemaking.	Definition of a PCB item.

LIMIT	AGENCY	DATE	EXPOSURE SCENARIO	APPLICABILITY & RESTRICTION
10 ppm	Consensus	1985	High public contact eg. residential site	Proposal Nationwide PCB spill Cleanup Standard.
25 ppm	Consensus	1985	Reduced Public access areas	Proposal Nationwide PCB Spill Cleanup Standard.
50 ppm	Consensus	1985	Restricted access areas, eg. electrical substances.	Proposal Nationwide PCB Spill Cleanup Standard.

their separation by cyclone and the reuse of the shot. Caution should be used in controlling the speed of use, as slower speeds result in friction heating, which in turn may vaporize PCB or alter its chemical structure.

Grinding or scarifying refers to the use of equipment with rotating carbide-tipped tools, arranged in a spiral configuration and used to remove up to 0.5 inches of concrete per pass. The cutting tool is rotated at high speed and mounted on a heavy tractor which moves it forward at a controlled speed. This technique is often used to grind surfaces off highways and bridge decks but has also been used in large industrial decontamination applications where PCB floor contamination has penetrated to depths of 1 to 2 inches.

Applications of the technique are limited in part by the size of the equipment required to do the job and, in some cases, by the weight of the grinding equipment in multistory buildings. Care must be taken in the latter situation to ensure that the technique does not have an adverse structural impact on the building as it is applied and concrete is removed. Some performance specifications for commonly used grinders are shown in Table 2 (Barbier and Chester, 1980).

Scabbling refers to the use of rows of small, high pressure impact pistons to break up and remove concrete. This technique, frequently used in the nuclear industry for decontamination, is applied in a similar fashion to grinding. Vacuum collection devices are often included as attachments but are relatively ineffective due to the high speed of application and the unpredictable size grading and deflection pattern involved in scabbling.

High pressure water cutting techniques have been refined for removal of concrete surfaces. One company developed a high pressure water device designed to cut through concrete up to several inches. The technique is intended primarily for the construction and bridge maintenance industry, but the system has been tested for contaminated concrete removal and has performed commendably. Large volumes of water are used, however, which presents the same water treatment and cross-contamination problems as pressure washing.

Deep Concrete Removal

Long-term open use of PCB in certain industrial applications has resulted in isolated cases of deep PCB penetration in concrete. PCB penetration in excess of 12 inches has been documented, sometimes saturating concrete all the way through to the subsoil. Similar penetration problems can be encountered where concrete integrity is lost through aging, particularly where cracking or spalling has occurred.

No particularly sophisticated techniques have been developed for deep concrete cleaning or treatment. Removal is typically accomplished using jackhammers, backhoes or other commonly used construction implements.

TABLE 2

COMPARISON OF VARIOUS CONCRETE SURFACE REMOVAL TECHNIQUES (REF. 3)

TECHNIQUES	LIMITATION	ESTIMATED RELATIVE SPEED AT WHICH A UNIT OF SURFACE AREA CAN BE REMOVED
Sand Blasting	Grit Adds to the Contamination	Slow
Dry Ice Blasting	Very Slow Penetration	Slow
Flame Spalling	Heat may cause Undesirable Chemical Reactions	Slow
Explosives	Genertes Moderate Quantities of Dust which Must be Controlled	Fast
Jack Hammer	Awkward to Use on Walls	Medium Fast
Impactor Powered by Air or Hydraulics	Limited to Large Accessible Facilities	Fast
Scrubber or Scabbler	Awkward to Use on Walls	Slow
Water Cannon Hand-held Modified 458 Magnum Rifle	Gun Powder Combustion Products are Produced	Slow (5-6 min/ft^2)
Rapid-Fire Model	Limited to Large Accessible Facilities	Slow (3-4 min/ft^2)
Concrete Spaller with 38-Pound Air Drill to Make Holes		
Hand-held		Medium Fast (50-60 sec/ft^2)
Semi-Automated on Platform		Medium Fast (35-40 sec/ft^2)
High-Pressure Water (40,000 to 60,000 psi)	Produces Contaminated Water	Fast (10-15 sec/ft^2)

In Situ Treatment

Reagent technology adapted from the PCB treatment industry is now being applied to decontaminate concrete surfaces. The technique has been successfully applied in Europe to several spill cleanups involving PCB, dibenzofuran (PCDF) and dibenzodioxin (PCDD). In all cases, the technique met or exceeded the performance of other removal techniques at a significantly lower cost. Table 3 illustrates one example of reagent effectiveness on concrete (Nobile, et al., 1986).

The liquid reagent, a polyethylene glycol-based mixture, is heated and applied to the surface (particularly floors) using a sprayer or brush. The reagent is then allowed to remain in place. (Second and third applications are common.) When it penetrates up to 2 inches, the liquid contacts the PCB and reacts to form non-PCB by-products. The technique is currently being tested in the United States and should soon be available for certain specific applications.

COATING TECHNOLOGY

Concrete sealants have often been used as a means of finishing industrial floors for spillproofing or wear resistance. These coating techniques have some applicability to PCB contamination situations on the principle that a sound coating material would at least temporarily inhibit human exposure in the work place. Several coating materials are commercially available and touted as specifically applicable to PCB decontamination.

In practice, because many ongoing decontamination projects involve industrial property transfer or long-term liability issues, leaving PCB in place even under a protective surface coating is not always acceptable to the buyer. Furthermore, coatings have found little acceptance with state agencies and the EPA as a permanent fix. Like capping a landfill, coating PCB contamination in concrete requires long-term maintenance and long-term responsibility for the user. For these reasons, surface coatings are not considered a decontamination technique per se but serve as a finishing technique following decontamination by other methods.

CHOOSING A DECONTAMINATION TECHNIQUE

Most significant decontamination projects require the selection and use of several techniques for different situations and depths of contamination in concrete. Typical evaluation criteria for the selection process include:

- Production/removal rates
- Effectiveness
- Potential for cross contamination
- Potential for enhanced penetration
- Waste treatment and disposal requirements
- Cost.

These factors are sometimes interrelated; .that is, the selection of one technique over another means a trade-off between, for example, waste disposal and productivity.

TABLE 3

COMPARISON OF ROAD CUTTING MACHINES FOR EQUAL
CUTTING DEPTH (1.25 cm) AND SPEED (720 m/h) (1980) (Ref. 4)

MANUFACTURER AND MODEL	CUTTING WIDTH (m)	AREA PER UNIT TIME (m²)	COST					COST PER TOTAL ($/h)	COST PER UNIT AREA ($/m)
			OWNERSHIP ($/h)	MAINTENANCE ($/h)	BITS ($/h)	LABOR ($/h)			
Dresser Galion Road Planner									
RP 12	0.314	226	15	11	19	18	63	0.28
RP 30/42	1.066	768	36	16	47	18	117	0.15
RP 60	1.524	1097	45	18	97	18	178	0.16
CMI Pavement Profiler									
PR-275-RT	2.032	1463	88	29	159	20	296	0.20
PR 525	2.642	1902	99	56	268	20	443	0.23
PR 750	3.78	2722	118	71	540	20	749	0.28
Barber-Greene Dynaplane									
RX-40	1.875	1350	90	90	207	75	462	0.34
RX-75	3.15	2268	146	105	348	75	709	0.31
RX-75	3.725	2682	149	115	411	75	750	0.28

Production/removal rates for some commonly used floor and wall concrete removal techniques are shown in Table 4. In general, the most productive techniques are also the crudest and generate the most waste. Application of concrete decontamination in a relatively sensitive environment is easier to control but will typically require much slower production equipment such as that used in shot blasting. For heavy industrial applications with large floor spaces and limited impediments to movement, grinding or high pressure water techniques can be employed (Woodyard, 1985).

Comparative effectiveness of selected cleaning techniques is shown in Table 5. When gauging the comparative success of these techniques on concrete, however, effectiveness alone is not a good evaluation criterion. Concrete decontamination situations typically involve surface cleaning or removal to a specified depth and are therefore not readily measured by percent removal of PCB. Proper characterization of the site is the key to the effectiveness of the removal technique, as it dictates what depth of removal is required.

The potential for cross contamination exists during concrete decontamination, as many of the removal techniques are dusty or wet and therefore have a high potenial for contamination of other clean surfaces. As noted above, the slower, better controlled techniques, such as shot blasting, tend to minimize cross contamination while the cruder large surface techniques, such as grinding or hydroblasting, generate the most waste and the most airborne particulate/aerosol.

More important, there is the potential for enhanced penetration during concrete decontamination if the appropriate techniques are not properly selected and applied. Haphoyard use of wet techniques in particular has been shown repeatedly to drive PCB contamination deeper into floors. In situations where surface sampling criteria are used to gauge effectiveness, this problem may go unnoticed and eventually result in contamination at lower depths. Wet techniques are to be avoided where possible for this reason.

Waste generation from decontamination operations has always been a concern arising not only from disposal costs but from the liability associated with waste treatment and disposal requirements governing landfilling of solid hazardous waste. The most prolific waste generators among the dry decontamination techniques described here include sandblasting and grinding. The most prolific wastewater generators are hydroblasting and pressure washing. Situations involving PCDFs and PCDDs provide an additional cause for waste disposal concern, as some of these materials cannot be readily landfilled or incinerated and must remain on site awaiting disposition. Table 4 lists some observed waste generation volumes for several of the more common decontamination techniques.

Cost is always an important criterion when selecting techniques for decontamination. However, it should be clear from the preceeding discussions that each particular situation provides only a limited number of competing technologies. Removing 2 inches of floor concrete in a large manufacturing facility, for example, forces the contractor to choose among grinding, scabbling or complete demolition. In general, the technique itself is not the principal determinant of cost, but rather the production rate and crew size balanced against waste treatment and disposal requirements. Techniques such as shot blasting are slower than grinding on a large scale but generate little waste and are much easier to control.

TABLE 4

LIQUID REAGENT DECONTAMINATION OF SURFACES
CONTAMINATED BY PCDFs AND PCDDs (REF. 5)

PCDFs AND PCDDs	SAMPLE A (Floor)		SAMPLE B (Wall)		SAMPLE C (Roof)	
	BEFORE (ng/m^2)	AFTER (ng/m^2)	BEFORE (ng/m^2)	AFTER (ng/m^2)	BEFORE (ng/m^2)	AFTER (ng/m^2)
2,3,7,8 TCDF	256	N.D.	29	N.D.	1060	174
TCDF	1169	27.4	132	N.D.	4841	362
PeCDF	678	13.4	77	N.D.	2809	92
HxCDF	426	N.D.	48	N.D.	1762	22
HpCDF	336	11.0	38	N.D.	1391	N.D.
OCDF	248	19.8	28	N.D.	1029	N.D.
tot. PCDF	2857	71.6	323	N.D.	11832	476
2,3,7,8 TCDD	N.D.	N.D.	N.D.	N.D.	33.1	N.D.
TCDD	24	N.D.	3	N.D.	100	N.D.
PeCDD	36	N.D.	4	N.D.	148	N.D.
HxCDD	45	N.D.	5	N.D.	187	N.D.
HpCDD	42	12.8	5	N.D.	174	N.D.
OCDD	35	114	4	70	143	N.D.
tot. PCDD	182	126.8	21	70	752	N.D.
tot. (PCDF+PCDD)	3039	198.4	344	70	12584	476

TABLE 5

APPROXIMATE PRODUCTION RATES FOR SELECTED
CONCRETE DECONTAMINATION TECHNIQUES

	PRODUCTION RATES (sq ft/min)	
TECHNIQUE	FLOORS	WALLS/CEILINGS
1. Grinding (a)	50	
2. Scarifying (a)	12	
3. Shotblasting	6	
4. High Pressure (a) Hydroblasting	4	8
5. Sandblasting		5
6. Grit Blasting (Metal)		5

(a) Assumes a 0.5 in. depth of cut in concrete.

TABLE 6

GROSS AND FINAL DECONTAMINATION TECHNIQUES AND THEIR EFFECTIVENESS (REF. 6)

CONTAMINATED SURFACE	DECON TECHNIQUES	REMOVAL EFFECTIVENESS (%) *
Concrete Floor+	• Scrubbing with detergent • Acid etching • Sandblasting/shotblasting	50 - 95 80 - 98 25 - 95
Unpainted cinder-block wall	• High pressure water spraying • Steam cleaning	60 - 90 90 - 99+
Galvanized painted ceiling and walls	• High pressure water spraying	80 - 99+
Painted steel walls, ceiling, floor & cabinets	• High pressure water spraying • Various methods	70 - 95 10 - 50

* Single Cleaning Pass
+ Assuming short term contact, no standing liquid

TABLE 7

APPROXIMATE WASTE TREATMENT AND DISPOSAL REQUIREMENTS FOR SELECTED CONCRETE DECONTAMINATION TECHNIQUES

		WASTE GENERATION	
	TECHNIQUES	SOLID WASTE (lb/sq ft)	WASTE WATER (gal/sq ft.)
1.	Grinding	0.5	
2.	Scarifying	0.5	
3.	Shot Blasting	0.5	

4.	High Pressure Hydroblasting
	  a) Walls
	  b) Floors
			0.3
			1.0

5.	Sandblasting	6	
6.	Grit Blasting (Metal)	0.5	

Because multiple techniques are often used on a given project, it is difficult to estimate the exact cost of applying a particular technique on a specified surface. In general, industrial building decontamination involving only concrete cleaning will cost around $1.00 to $2.00 per square foot of surface. Concrete removal to greater depths will increase the cost significantly to a point where the concrete must be removed in its entirety, at which point the cost will either decline or escalate dramatically.

CONCLUSIONS

The technology for PCB decontamination of concrete includes a wide variety of techniques with specific applications in industry. Each decontamination project typically involves several different depths of PCB contamination and a variety of concrete ages and conditions.

The selection and application of decontamination technology improve with each project. Several techniques are on the development horizon that could have a significant impact on the industry today.

Because most decontamination equipment is applied from other industries, considerations such as cross contamination and waste generation have not been factored into the design and are not compatible with PCB waste management principles. Care must be taken in the selection of decontamination technologies to ensure that waste volume is minimized where possible and cross contamination does not occur or can be controlled to acceptable levels.

The extent of concrete contamination is not well defined in most circumstances and never will be completely understood because of its very nature. Selection of the right decontamination technique must be accompanied by the know-how required to implement it so that waste and cross contamination may be controlled while effectiveness and production are maximized.

REFERENCES

Barbier, M. M. and C. V. Chester, May 1980, "Decontamination of Large Horizontal Surfaces Outdoors," presented at the USDOE Concrete Decontamination Workshop, Richland, Washington.

International Technology Corporation, 1986, "Case Studies of PCB Transformer Fires" (Draft), prepared for the Electric Power Research Institute, Palo Alto, California, Contract RP1263-20.

Nobile, G., W. Tumiatti, and P. Tundo, May 1986, "In Situ Decontamination and Chemical Degradation of PCDFs and PCDDs Coming From Thermal Oxidation of PCBs," presented at the American Chemical Society Division of Environmental Chemistry Symposium, New York.

Woodyard, J. P., Dec. 1985, "State-of-the-Art Decontamination Technology for PCB Spills and Fires," presented at HAZMAT 85 West, Long Beach, California.

Woodyard, J. P. and R. L. Wade, May 1986, "Sampling and Decontamination Methods for Buildings and Equipment Contaminated With Polychlorinated Dibenzodioxins," presented at the American Chemical Society Division of Environmental Chemistry Symposium, New York.

THE RETROFILL AND RECLASSIFICATION

OF POLYCHLORINATED BIPHENYL TRANSFORMERS

Ian Webber

Envirocon Systems Inc.
57 Sandlewood Drive
Getzville
NY 14068

INTRODUCTION

The severe health effects observed in the Japanese 'Yusho' incident of 1968 were attributed to the ingestion of polychlorinated biphenyls (PCBs). At that time, the forefront of analytical chemistry was represented by the determination of trace components at the parts per million (ppm) concentration level. It was not until about ten years later that analytical methodology was able to detect polychlorinated dibenzofurans (PCDFs) and polychlorinated dibenzodioxins (PCDDs) at concentrations of 10 parts per billion (ppb) or less in the presence of PCBs. The significance of the determinations lies in the assessment of risk to human populations exposed to undegraded PCBs and to mixtures of chemically similar compounds of concern derived from uncontrolled reactions such as might occur when a PCB filled transformer undergoes eventful failure.

A brief discussion is included in this paper of the realistic assessment of the fire hazards of PCB replacement liquids in electrical transformers.

The costs of retrofilling PCB transformers are contrasted with the costs of retrofitting.

Many attempts have been made to reclassify PCB impregnated transformers but only now has a process been proven successful. This paper explains some of the reasons why conventional retrofilling operations are usually insufficient to achieve reclassification and, more importantly, why the reclassified status of a conventionally retrofilled transformer is not maintained.

Experiments have been performed to assess the diffusion coefficients of PCBs and chlorobenzenes through kraft paper into the bulk retrofill oil. A mathematical model is presented which uses the experimental data to describe the leaching effect of PCBs from porous transformer components. The calculations have been used to derive a correlation between the extent of the initial cleaning and the time in days for the leaching of PCBs into the bulk oil to reach 500 ppm PCB. The theoretical model is substantiated from experimental data on the conventional retrofill of a 270 gal., 750 KVA askarel transformer. Experimental data are also presented to show that a PCB impregnated transformer can be reclassified in a single operation and also that it will retain its reclassified status for the remainder of its service life.

The retrofill of mineral oil transformers defined as PCB units with greater than 500 ppm PCB is a trivial problem relative to the reclassification of askarel units. The reclassification of this type of transformer can be achieved with conventional solvent cleaning equipment.

The commonly used method of reclassification by repeated draining and flushing yields three or more times the volume of the transformer of PCB contaminated fluids for disposal. At 2000 ppm PCB and higher concentrations, chemical decontamination processes which can be used with the transformer on.line do not compete cost-effectively with incineration. The use of a vapor cleaning process to minimize the quantity of fluid for disposal then becomes cost-effective. The process is a physical requirement when the PCB concentration of the residual mineral oil is about 2000 ppm PCB.

COMPOUNDS OF CONCERN

In 1968 a group of workers at Kyushu University in nothern Japan determined that a group of patients with acneiform eruptions similar to chloracne had ingested contaminated rice oil used in cooking. The symptoms of the poisoning suggested that it was due to organochlorine compounds. The word 'Yusho' or 'rice oil' disease is used to describe the incident. The number of recognized victims over a period of about ten years rose to a total of approximately 1800 people.

The rice oil contamination was determined to be Kanechlor 400 which was produced as an approximately 48% chlorine substituted biphenyl. The fluid was used by the rice oil manufacturer as a fire hazard reduction measure in an indoor heat exchanger.

The PCB concentration in Yusho oil was found to be about 1000 ppm PCB (Kuratsune et al. 1972). As analytical capabilities increased there has been a parallel increase in the speculation of the relative toxicities of other causative agents such as polychlorinated dibenzofurans (PCDFs) and polychlorinated quaterphenyls (PCQs). (Masuda and Yoshimura 1984). Original estimates of the PCB contamination were in the range of 2000-3000 ppm but were found to be in error primarily because of the analytical interference presented by PCQs.

The early analytical method applied to the quantitation of PCBs in oils uses a calibration technique developed by Webb and McCall in 1973. The method is based upon the detector response comparison obtained for a commercial PCB mixture with that of the sample and uses packed column gas chromatography with, typically, an electron capture detector (see ASTM method D-4059). Slight variations in the aroclor manufacturing process, or chemical alteration of the analyte PCB, causes problems with the quantitation. Reasonable results can be achieved, however, in the analysis of insulating oils where the PCB has not been degraded. More sophisticated analytical methods have been published by Albro et al. (1979), Lao et al. (1976), Sherma et al. (1975) and Stalling (1979).

A poisoning very similar to the Japanese Yusho incident occurred in Taiwan in 1978-1979 and is known as the 'Yu-Cheng' incident. In this case also, a leaking heat exchanger contaminated rice oil during the manufacturing process with a mixture of Kanechlor 400 (48% Cl) and Kanechlor 500 (54% Cl). 1,843 cases of poisonings were reported during the period 1978–1979 (Masuda and Yoshimura 1984μ Kunita et al. 1984μ Chen et al. 1984). The PCB concentration in the oil was less in this incident than in Yusho but the higher dose rates to the victims resulted in approximately the same total dose.

Hsu et al. (1984) have estimated that the minimum effective dose of PCBs in the Yu-Cheng incident was 0.3 to 0.5 g. The ratio of PCDFs to PCBs in the Yu-Cheng rice oil was found by Chen et al. (1984) and Kunita et al. (1984) to be approximately 1:300 and therefore the minimum effective dose for PCDFs is of the order of 1 mg.

The assessment of potential toxicity is exacerbated by the fact that not all PCB, PCDF, PCDD, etc., isomers are equally toxic. It is therefore not sufficient to analyze for the presence of a class of compounds but also necessary to quantitate the biologically active isomers within a class.

There are 209 PCB isomers contained within the 10 homologous series of PCBs. The series result from the number of chlorine atoms substituted on the biphenyl rings. A commercial aroclor mixture may contain 60 or more different isomers.

PCBs are absorbed in the fatty tissues of humans. The ultimate fate of the PCB isomers depends upon their pharmacokinetics so that the analysis of isomers remaining in a tissue may show little resemblance to the isomers originally ingested. Less chlorinated PCBs tend to be metabolized more readily than their higher chlorinated congeners and therefore aroclors such as aroclor 1260 or 1254 tend to bioconcentrate more than aroclor 1242.

There have been several studies which have concluded that past exposure to PCBs is associated with excess mortality or morbidity from cancer in human populations (Bertazzi et al. 1981: Humphrey 1983: Unger et al. 1982). However, the health effects in occupationally exposed workers are typically complicated by such factors as incomplete exposure information, job turnover and low chemical specificity.

The most prominent and consistent health effect associated with exposure to PCBs is chloracne and dermatitis. These effects usually clear up quickly and, if they were the only effects, would reduce the need for concern.

It is beyond the scope of this paper to investigate all the health effects associated with PCBs but there is little doubt that PCBs are compounds of concern. The concentration at which this occurs is much higher than that of many other classes of chlorinated aromatic compounds. The risk associated with exposure to PCBs can be separated into a consideration of the biological activity of PCB isomers compounds contained in the PCBs as impurities and the biological activity and probability of formation of degradation products.

Estimates of the dosage of toxic compounds received by victims of the Yusho and Yu-Cheng poisonings are dependent upon the chemical specificity of the analytical methods used. There is no doubt that the PCB involved had undergone partial oxidation and consequently, the overall toxicity of the fluid was the combined effect of compounds of concern.

In 1970 Vos et al. showed a correlation between the toxic effects of European PCBs and the concentration levels of PCDFs. The relative concentrations of the PCDF isomers present in Yusho oil and in two samples of other used heat exchanger PCBs (Kanechlor KC400 and Mitsubishi – Monsanto T 1248) were found to be strikingly similar. This fact underscores the work of Kunita et al. (1984) who have concluded that not only are PCDFs orders of magnitude more toxic than PCBs but claimed also that there is a synergistic action between them. The evidence suggests that PCDFs were a major causative agent in the Yusho and Yu–Cheng incidents relative to PCBs.

It is not yet possible to estimate the sensitivity of humans to the toxic effects of PCDDs. It has been suggested that humans are resistant to the toxic effects of PCDDs. The supposition is not presently verifiable because of the lack of characterization of human exposure. There is no doubt, however, that the 2,3,7,8-tetrachloro-DD (2,3,7,8-TCDD) isomer is a potent toxin which has caused impaired reproduction and/or birth defects in every species tested and has been described as the most toxic compound known to man.

In addition to PCDFs and PCDDs mentioned above there are several classes of polychlorinated aromatic hydrocarbons which may be contained as impurities or degradation products of PCBs. These classes include polychlorinated terphenyls (PCTs), polychlorinated quaterphenyls (PCQs), polychlorinated quaterphenyl ethers (PCQEs), polychlorinated naphthalenes (PCNs), polychlorinated biphenylenes (PCPs), polychlorinated pyrenes (PCPYs) and polychlorinated chrysenes (PCCYs).

Polychlorinated terphenyls have been found (Allen et al. 1973) to have qualitatively similar enzyme inducing effects to those of PCBs but appear to be less potent (Goldstein 1980). Mixtures of PCBs and PCTs have been marketed but were not widely used.

Polychlorinated quaterphenyls were found as contaminants of Yusho and Yu–Cheng oils and for a long time were an unrecognized interference in the quantitation of PCB contamination in these incidents. Studies by Hori et al. 1982 and Kunita et al. 1984 show that the toxic effects of PCQs are similar to PCBs.

Polychlorinated quaterphenyl ethers (PCQEs) have been identified in Yusho and Yu–Cheng oils. This class of compounds does not seem to have been tested for toxicity but structure/activity relationships do not predict that they should be particularly toxic. Nevertheless, cyclization to a phenyl substituted dibenzofuran possibly could yield a toxic product. The likelihood of the cyclization reaction provides a measure for the concern to be accorded the presence of ether precursors.

Polychlorinated napthalenes (PCNs) have been characterized in commercial PCBs and probably resulted from the presence of napthalene as an impurity in the biphenyl raw material used for the manufacturing process. Typical concentrations in PCB fluids are less than 600 ppm. Results obtained by Kimbrough (1981) and Goldstein (1980) indicated that it is unlikely that PCNs would add significantly to the toxicity of PCBs at the concentration levels typically observed.

Polychlorinated biphenylenes (PCPs), because of their structure, are expected to be as toxic as the correspondingly substituted PCDDs. According to Poland and Glover (1977), 2,3,6,7-tetrachlorobiphenylene is approximately as potent as 2,3,7,8-TCDD. PCPs have been found to result from the reductive conditions which occur in the early stages of askarel transformer fires and are therefore compounds of concern (Smith et al. 1982: Rappe et al. 1982: Williams et al. 1984).

Polychlorinated pyrenes (PCPYs) and polychlorinated chrysenes (PCCYs) have been identified as components of the soot from an askarel transformer fire (Rappe et al. 1983: Williams et al.`1984). The toxicities of these classes of compounds have not been determined but their concentrations are so low that they probably do not add significantly to the overall toxicity of the degraded PCB.

FORMATION OF COMPOUNDS OF CONCERN

Compounds of concern can be contained in PCB fluids for a variety of reasons. For example, they may be formed from impurities in the feedstock used to manufacture PCBs, or from the cyclization of PCBs induced by heat, or even from phenolic or ether precursors.

Polychlorinated terphenyls, quaterphenyls and napthalenes could result from feedstock contamination by traces of the aromatic hydrocarbons before the mixture was chlorinated. PCNs have been identified as a pyrolysis product from an askarel transformer fire. Their formation has been explained (Buser, 1979) in a laboratory study by invoking the formation of benzyne intermediates or the re-arrangement of intermediates formed between an ortho-chloro phenyl radical with a chlorobenzene.

The overall effect of radical reactions on the product distribution during askarel degradation will be effected by both temperature and the availability of oxygen.

Chittim et al. 1979 have analyzed unused North American askarels and found that they contained less than 0.05 ppm of PCDFs. It is therefore of concern that these workers found a correlation between the concentration of TCDFs in used askarel with the length of time the fluid had been in service. The small number of samples tested was insufficient to obtain a definite correlation between TCDF concentration and transformer loading. While the results indicated that transformer load is probably a major factor in the formation of PCDFs, the effects of discharging or arcing appeared to be negligible.

The Power Transformer Guide ANSI-C57.92-1981 lists a maximum top oil limit of 110°C which may occur when the maximum hot spot conductor temperature is at 180°C due to short term loading. The Distribution Transformer Guide ANSI C57.91-1974 lists a maximum top oil temperature of 120°C which may occur at a maximum conductor hot spot temperature of 200°C.

Maximum fluid temperatures in a transformer could exceed values in the overload guide if units are subjected to more severe loading. Units with extreme overloads would be expected to overheat and fail. The most probable cause of eventful transformer failure is violent rupture (Doble, 1982) and the quantities of any compounds of concern which may be produced are determined by the availability of oxygen in an oxygen depleted, high temperature reaction zone of short duration.

Buser (1979) has shown that pyrolysis of chlorobenzenes at 600°C in the presence of excess air yielded about 1% yield of tetra- to octa-CDFs and tetra- to octa-CDDs. It has also been shown (Buser and Rappe, 1979) that the pyrolysis of PCBs occurs intramolecularly by four alternative reaction routes to yield different isomeric PCDF products.

The thermal degradation of PCBs can result in a complex set of reactions which may produce compounds of concern under uncontrolled conditions. Temperature and residence time relationships have been extensively studied to establish the conditions necessary for satisfactory destruction (Mescher et al. 1978). It was found, for example, that with a 1s residence time, most PCB decomposition occurred in a temperature range between 640°C and 740°C. Commercial incineration equipment for the destruction of PCBs is therefore designed so that the energy input to disrupt the molecule is made available either by supplying a very high temperature or a long residence time at a relatively low temperature. Several types of incinerator are found useful for this purpose and include rotary kilns, high temperature fluid wall reactors, plasma pyrolysis units, circulating bed combustors etc.

Uncontrolled reactions which lead to the formation of chlorophenols as a side reaction product have the potential to generate PCDFs and/or PCDDs in the following ways:

(1) The dimerization of chlorophenates:
(2) The cyclization of polychlorinated diphenyl ethers:
(3) The cyclization of polychlorinated phenoxy phenols.

The Seveso accident in 1976 was the most recent of six similar occurrences in which uncontrolled reaction conditions caused the dimerization of trichlorophenol produced from tetrachlorobenzene and sodium hydroxide in ethylene glycol to yield PCDDs. In the case of liquid systems, reactants are retained in the reaction zone for periods of time which are long compared to the time required for the formation of product. In a gaseous reaction, on the other hand, such as in a flame, the reaction zone is relatively very short-lived and the yield of product is therefore less. The importance of even statistically unlikely reactions lies in the toxicity of the products.

The pyrolysis of PCDPEs follows two competitive reaction pathways viz., reductive dechlorination or ring closure to PCDFs.

The cyclization of polychlorinated phenoxy phenols is also a bimolecular reaction which yields PCDDs under the influence of heat.

PCB DEGRADATION IN BUILDING FIRES

The formation of pyrolysis products will ultimately depend upon factors such as the following: (i) the time that the askarel mixture is at a temperature which allows a reaction yield of significance: (ii) the volume and surface area of droplets of askarel emitted in an eventful failure: (iii) the availability of oxygen and (iv) the effect of soot particles on the dissipation of heat and the availability of oxygen.

Real-world experience has shown that the eventful failure of electrical equipment is frequently catastrophic. In particular, there is no doubt that transformer fires do produce askarel oxidation products in significant quantities. Studies have indicated that the breakdown of askarel transformer dielectric systems due to multi-stress aging may produce toxic products at concentrations of concern. The yield and distribution of products can be expected to depend upon the availability of oxygen. The major quantity of toxic products will undoubtedly be derived after the failure of the transformer has resulted in a fire situation.

When a transformer is retrofilled to <500 ppm PCB, one would expect that the rate of conversion of PCBs would be very much less than in the case of an askarel fire. It becomes increasingly less probable that a large enough quantity of pyrolysis products would be produced as the concentration of PCBs and chlorobenzenes is reduced. These qualitative expectations were realized when a fire occurred in the basement transformer vault of the Electrical Engineering laboratory of the University of Manitoba in 1982. The six transformers involved in the fire contained mineral oil contaminated with 250 ppm PCB. An electrical failure caused one of the transformers to arc and catch fire. All six transformers were damaged. During fire—fighting attempts a transformer exploded and distributed PCB contaminated soot onto the walls. No PCDFs were found. PCDDs, which are typically formed at an order of magnitude less concentration than PCDFs, were not detected either.

A fire occurred on February 5, 1981 in the Binghamton State Office Building in Binghamton, New York which involved a PCB filled transformer located in the basement. The building was closed and sealed shortly after the fire and remains closed today.

PCBs and oxidation products were distributed throughout the 18—story structure via two ventilation shafts. Analysis of soot samples revealed high concentrations of PCBs, PCDFs, PCDDs and also other classes of toxic chlorinated aromatic compounds.

The cost of the cleaning operation is estimated at $20,000,000 and exceeds the cost of construction. The expense of disposing of the building and replacing it was estimated at about $100,000,000.

Approximately $1.2 billion in law suits have been entered against the State in this incident.

A PCB transformer fire has also occurred in the transformer vault at One Market Plaza, San Francisco. The vault contained three transformers. Soot laden smoke issued from the sidewalk grating adjacent to the building for about three hours and contaminated the switch gear room and some areas of the adjacent parking garage. Analysis of the soot collected in the vault showed the presence of the same classes of compounds as found in the Binghamton incident. The conversion efficiency of PCBs to PCDFs and PCDDs seemed to have occurred with approximately the same efficiency as in Binghamton.

Estimates of clean—up costs are between $15,000,000 and $20,000,000.

In September 1983, a fire occurred in a transformer vault under the plaza on the same block as the First National Bank building in Chicago. Smoke issued from a sidewalk grating for about 45 min. but, fortunately, significant PCB contamination was limited to the vault.

Other PCB transformer fires have occurred in Miami and Tulsa.

The known costs of the fire related incidents described above has led the EPA to assess an average clean—up cost per incident of $20,000,000.

The Toxic Substances Control Act stemmed from the Yusho incident. Section 6(e) of TSCA requires proper disposal of PCBs, and prohibits the manufacture, processing, distribution in commerce, and use of PCBs. Under section 6(e)(2) of TSCA, the EPA allows the use of PCBs in 'a totally enclosed manner' such as transformers, capacitors and electromagnets based upon considerations of cost and risk reduction.

However, when the August 1982 rule was promulgated it was based upon the assumption that the principal route of release of PCB-containing dielectric fluid into the environment was from leaks and spills. Since that time, it has been found that fires involving transformers occur frequently enough to present a significant risk and this has led the EPA to reconsider the August 1982 rule and place the following additional restrictions and conditions on the use of PCB transformers (40 CFR Part 761: 50 FR 29170) July, 1985:

(1) High secondary voltage PCB transformers (480 V and above, including 480/277 V) configured in a network fashion and used in or near commercial buildings must be removed from use, placed into storage or disposal, disposed, **or reclassified to PCB contaminated or non-PCB status by Oct. 1, 1990.**

(2) PCB transformers can no longer be installed in commercial buildings after Oct. 1, 1985.

(3) PCB transformers used in or near commercial buildings (other than high secondary voltage network PCB transformers) must be equipped with enhanced electrical protection, by Oct. 1, 1990, to avoid failures and fires from sustained electrical faults.

(4) All transformers must be registered with appropriate emergency response personnel and with building owners by Dec. 1, 1985.

(5) All PCB transformer locations must be cleared of stored combustible materials by Dec. 1, 1985.

(6) All PCB transformer fire-related incidents must be immediately reported to the National Response Center, and measures must be taken as soon as practically and safely possible to contain potential releases of PCBs and incomplete combustion products to waterways.

EPA defines commercial buildings to include all types of buildings other than industrial facilities and would include locations such as office buildings, shopping centers, hospitals, and colleges. A PCB transformer located in or near a commercial building is located on the roof of, attached to the exterior wall of, in the parking area of, or within 30 meters of a commercial building.

The EPA has had to consider both the benefits of PCBs as well as the availability of substitute materials balanced against the costs of regulatory control measures. It was concluded that the removal or retrofill of PCB transformers is both the most effective and the most costly measure for reducing the frequency of serious transformer fires. It was suggested that a less costly, but also less effective, alternative could be represented by providing better electrical protection of the equipment. The effectiveness of increased electrical protection is expected to approach that of phaseout/retrofill but EPA recognizes that electrical protective devices are also subject to malfunction and that PCB transformer fires can result from less common mechanisms of failure.

At this point it would be useful to clarify what is meant by the retrofill and reclassification of PCB transformers.

The retrofill of a PCB filled transformer involves the replacement of the original PCB dielectric fluid with a substitute oil. The main advantage of retrofilling is that an owner's liability is minimized at minimum cost. In addition, record keeping and reporting requirements are reduced or eliminated and servicing is allowed.

There are three categories of PCB transformers:

* A PCB transformer is defined as one which contains more than 500 ppm PCB.

* A PCB contaminated transformer is one which contains PCBs in the concentration range 50 – 500 ppm PCB.

* A non-PCB transformer contains less than 50 ppm PCB.

If the flushed carcas is buried in an approved chemical waste landfill there is the possibility of long term liability associated with PCB leaching. Alternatively, carcases can be cleaned to less than 10 μg PCB/100 cm^2 and the clean metal returned to the secondary metals market.

PCB transformers may not receive servicing which requires the removal of the core/coil assembly. Inspections must be made for leaks and records kept. When the unit is to be replaced, the fluid must be burned in an approved incinerator and the transformer carcas either incinerated or flushed.

PCB contaminated transformers (50–500 ppm PCB) do not need to be inspected for leaks and may be repaired or serviced as necessary. EPA reaffirmed its August 1982 determination that the continued use of PCB contaminated transformers does not present unreasonable risks to public health and the environment.

Non-PCB transformers, containing less than 50 ppm PCB, may be used with almost no restrictions since the unit is then not under TSCA purview.

The May 1979 PCB ban rule prohibits rebuilding of PCB transformers but does allow for their reclassification. Under the provisions for reclassification the equipment must be put back into service for three months before the bulk oil is tested for its PCB concentration. The final rule clarifies the definition of in-service use for transformers by specifying a minimum dielectric fluid temperature of 50°C to correspond to a condition of light electrical loading.

If a unit is retained as a PCB transformer an immediate cash outlay is avoided but this might not be the lowest-cost option in the long term. For example, if the unit should require servicing at some time in the future it will have, instead, to be scrapped. If tests showed that the transformer was about to fail it would not be possible to retrofill it and service the unit as a PCB contaminated or PCB free unit until it had been set back in operation for 90 days after the retrofill. This may be neither possible nor feasible, depending upon the condition of the transformer.

If an owner makes the decision not to continue to operate a unit as a PCB transformer he has then to choose between replacement with a new unit or retrofill of the old one. The age and performance history of the transformer become factors as well as how long the unit is likely to be needed.

The following cost elements enter the business alternatives of replace vs. retrofill:

Replacement

* Basic cost of transformer and dielectric coolant.
* Miscellaneous fittings — stress cones, etc..
* Possible rental of mobile substation during changeover.
* Crane rental.
* Labor (probably at overtime rates).
* PCB packaging, transport and disposal.
* Purchase cost of flushing solvent to prepare transformer carcass for landfill disposal.
* Packaging, transport and disposal of flushing solvent.
* Transport to and disposal in EPA–approved landfill for transformer carcass.

Retrofill

* An initial, fundamental decision must be made concerning whether to change the classification of the transformer to 'PCB contaminated' ie., <500 ppm PCB, or to <50 ppm PCB and remove the unit from TSCA purview.
* Dielectric fluid costs.
* Solvent flushing fluid, depending upon the retrofill method.
* Transport of oil, solvent and materials to and from the site.
* Gaskets, parts, hosing, fittings etc.
* PCB disposal, solvent disposal.
* Downtime 4h to 48h, depending upon the level of decontamination required and the process used.
* Repeated electrical outages, depending upon the retrofill process chosen.
* PCB–in–oil analyses, dielectric fluid properties report.

The following cost comparison contrasts the options of retrofill or replacement of askarel filled transformers. The comparison is complicated by the many variables which are involved and therefore a range of costs have been considered as they apply to two different sizes of PCB filled transformers.

Transformer Description

	Transformer 1	Transformer 2
kVA	500	1500
Location	Indoors	Indoors
Primary Voltage (110kV BIL)	12470Δ	13800Y/7970
Secondary Voltage	480Y/277	480Y/277
°C Rise	65°C	55/65°C
Oil Cooled	OA	OA
Winding Temperature Indicator	Yes	No
HV Connection	Throat	Throat
LV Connection	Throat	Throat
Pressure Relief Device	Cover mounted	Cover mounted
Sudden Pressure Relay	Cover mounted	Cover mounted
Taps	2 ± 2^1/^{2}%	2 ± 2^1/^{2}%
Dimensions	4' x 6' x 6'	5' x 7' x7'
# of Gallons Oil	190	225

RETROFILL COSTS ($):

	Transformer 1	Transformer 2
* Askarel removal, vapor phase degreasing, Webber reclassification process	8,000	13,250
* Transportation	1,000	1,000
* Food/Lodging	500	500
* Oil	1,906	2,257
* Testing	600	600
* Supplies	1,500	1,500
* Equipment lease	133	133
* Insurance	200	200
* Contract coordination	1,459	1,459
* Askarel disposal	2,000	2,400
* Transportation charges of PCB waste disposal	1,000	1,000
* Miscellaneous	1,000	1,000
Sub-Total:	$19,300	$25,300
* Inspection savings	- 800	- 800
TOTAL	$18,500	$24,500

TRANSFORMER REPLACEMENT ($):

	Transformer 1	Transformer 2
* Replacement cost	$14,376	$24,734
* Change out flexible leads, parts, etc.	1,320	1,650
* Labour of transformer changeout	2,240	2,240
* Crane rental	1,200	1,200
* Disposal of PCB liquid waste	1,600	2,000
* Tank and winding solvent	855	1,000
* Flushing solvent disposal	1,600	2,000
* Labour of flushing PCB transformer	420	420
* Disposal of PCB transformer carcass	2,880	4,900
* Transportation	1,500	1,500
* Fluid transportation for disposal	1,000	1,000
* Cost of drums and drum disposal	500	640
* Protective clothing, miscellaneous supplies	1,000	1,000
TOTAL:	$30,500	$44,284

COST SAVINGS:	Transformer 1	Transformer 2
* Transformer Replacement	$30,500	$44,300
* Transformer Retrofill	18,500	24,500
COST SAVINGS	$12,000	$19,800
% COST SAVINGS	39	45

EPA has completed an analysis of the costs of retrofilling PCB transformers to reduce the PCB concentration to below 500 ppm PCB and has estimated that the cost will vary in the range from $15,505 for a 50 KVA transformer to $32,034 for a 3,000 KVA transformer. The example above is for an 'average' or 'typical' size transformer of about 233 gal. oil capacity and shows that the cost savings over replacement will be about $20,000 per unit. Clearly, the larger and more inaccessible a transformer, the more cost effective retrofilling becomes.

THE VIABILITY OF PCB REPLACEMENT FLUIDS

The need to reduce or eliminate PCBs from the dielectric fluid of transformers is partly because of the generation of compounds of concern in a fire incident. However, it is also important to consider that the original reason for the use of transformer askarels was because they were intended to provide a measure of fire safety.

Fire hazard factors for replacement askarel transformer fluids are neither well defined nor easily estimated, but include:

* the ignition sources available:
* the availability of fire protection systems:
* the ease of ignition of the transformer:
* the close environment of the transformer:
* the effects of burning liquid on the surroundings:
* the depletion of oxygen by burning liquid:
* the smoke evolution from burning liquid:
* the toxicity of combustion products:
* the temperature at which the liquid is used.

The test methodology used to describe a system as complicated as a real transformer fire is clearly in its infancy and while the factors which affect fire hazards can be listed in broad terms, their interdependent description in an overall assessment is difficult using the rudimentary test methods presently available. More importantly, the interpretation of such results to provide a quantitative measure of the overall fire hazard should be treated with caution.

Ideally, a fire hazard assessment should be predicated on incidence studies in actual usage. Such an assessment is difficult for 'less flammable' fluids, ie., fluids with a fire point above 300°C, because of the lack of statistics on fires in which these fluids are involved.

Section 450-23 of the National Electrical Code specified in the past that the controlling provision for fluids used in indoor transformer applications should be that the insulating liquid have a 300°C fire point. The required temperature has some validity in that it is known that flaming will not persist in a wood slab not subjected to supplemental heating unless the average temperature within the slab is greater than about 320°C.

In the absence of primary current limiting fusing, the intensity and duration of a high energy arc are limited by the recovery capability of the liquid. When there is sufficient energy to produce an explosion, liquid aerosol and gaseous decomposition products are expelled from the transformer as a hot plume. The magnitude and probability of potential loss will be partially dependent upon whether the plume ignites. Subsequently, the effect of the transformer failure will depend upon the ignitability of the fluid and the materials surrounding the transformer.

A report was issued by the National Electrical Manufacturers Association (NEMA) in 1980 concerned with 'Research on Fire Safety Test Methods and Performance Criteria for Transformers Containing PCB Replacement Fluids'. The scenario used by NEMA to describe an eventful failure is as follows:

1. Incipient fault induction period.
2. Growth of incipient fault to large arc.
3. Failure of electrical protection devices to remove the transformer from the line before tank rupture.
4. Tank rupture with scattering fluid, gaseous decomposition products, solid insulation, steel components and molten conductor leading to ignition of nearby combustible building materials and furnishings by the transformer.
5. Scattered materials in static condition within minutes after tank rupture resulting in ignition of combustible building materials and furnishings.

NEMA has recognized that the principal underlying cause of transformer failures involves insulation breakdowns. A typical failure was described as beginning as an insulation failure in the high voltage winding and starts as a low current, high impedance fault. This turn to turn or layer to layer fault progresses rapidly, involving additional turns and layers. The coil impedance decreases and the fault current increases. An alternate failure model begins with a high impedance path to ground along the surface of insulation (tracking).

The following scenario was formulated to serve as the basis for a worst case failure:

1. Due to malfunction, the electrical breakers and/or fuses which are directly associated with the particular transformer are inoperative. The sustained presence of a severe overload current therefore goes undetected, permitting the insulating oil to overheat significantly.
2. As a consequence of the overload, the average oil temperature in the tank reaches 135°C. (Note: the NEMA study accepted this temperature as 'an arbitrary but reasonable level to assume under the circumstances').
3. Insulation deterioration occurs, causing charring or tracking of the high voltage insulation. A fault to ground develops through the train of bubbles or along the insulation surface to ground. The tracking fault then rapidly flashes into a long, high current arc.
4. Once the long arc has developed the sequence involves the following:
    a. The long arc rapidly decomposes the surrounding oil into gases.
    b. A high pressure (several hundred psi) gas bubble develops around the arc under the oil.
    c. The pressurized bubble rapidly accelerates the oil upward towards the gas space and also subjects the tank walls below liquid level to a severe overpressure.
    d. Depending on the tank geometry, failure occurs either near the bottom, below the oil surface or at the top by ejection of the cover.
5. Ignition of the hot oil spray results from one or more of the following sources:
    a. The hot solid particles of insulation and conductor which are produced by the arcing fault.
    b. The arc drawn by the ejected bushing.
    c. The hot gas bubble of arc decomposed oil which, at a temperature of about 3000°C initially drove the oil out of the tank, is itself ejected.

The theory of pressure phenomena due to arcing in liquid filled transformers has been studied and relationships derived between arcing and tank pressure (EPRI project 325, 1976). Tests were conducted on different types of arcs to corroborate the results . For example, current flowing through an expulsion fuse produces a higher arc-voltage gradient than does the same current in a free arc in oil. Higher voltage resulted in higher arc energy, and was accompanied by higher peak pressure in the tank. Faults contained within the coil's windings were found to produce less pressure in the transformer than either an open arc or an expulsion fuse. In particular, short length winding faults were found to be less severe than one inch arcs drawn directly in oil. Peak values of below-oil pressures were observed when the melting of a fuse wire initiated the arc, since this resulted in a high, near instantaneous rise in arc current. The maximum pressures developed under the oil for fuse initiated arcs were found to be very high, in the range of 20-30 atmospheres.

Electrical failure of cylindrical 10 kVA tanks has been reported to exceed 100 kW-sec. but for rectangular tanks this value increases to 800-1400 kW-sec. Values for cylindrical tanks are found to grow rapidly with increasing tank size. Therefore, eventful failure is much more likely in small cylindrical tanks.

The real world test parameters discussed above have been incorporated into the fire hazard assessment model used by UL (Webber 1983) to determine the compliance of transformer fluids with section 450.23 of the National Electrical Code. In order to be compliant with the code, transformers insulated with less-flammable liquids are permitted to be installed without a vault in Type I and Type II buildings of approved noncombustible materials in areas in which no combustible materials are stored, provided there is a liquid confinement area, the liquid has a fire point of not less than 300°C and the installation complies with all restrictions provided for in the listing of the liquid. UL has identified the need for pressure relief devices and current limiting fusing to limit the effect of possible high current arcing faults.

A report describing the predominant transformer failure mode has been presented by D.A. Duckett, (1975). The purpose of this early testing was to determine the flammability of dielectric liquids after an explosion. Approximately four gallons of each of the fluids under test were placed in separate transformer tanks and preheated to 150°C. Each container had internal electrodes designed to force the arc upward into the gas space. Although the temperature at the point of arcing was several thousand degrees, the duration was only for a few cycles and therefore the temperature of the fluid in a container the size of even a small distribution transformer cannot be significantly affected. The 150°C used in the test was probably higher than would reasonably be expected to occur and therefore the same tests were also run at 120°C. It was found that a fireball was not produced at the lower fluid temperature.

When the test conditions were made sufficiently severe to expel liquid and gas from the transformer at high temperature, a high molecular weight paraffinic hydrocarbon oil produced a smaller fireball, lower fireball temperature and shorter fire residence time than any of the other fluids tested, including silicone dielectric oils.

Catastrophic failure test methods such as that described above are difficult to standardize because of the many factors which contribute to the results. Nevertheless, the scenario used is substantially more realistic than many other fire tests applied to less flammable fluids. For example, ignitability tests to determine whether a fluid would contribute to a fire showed that, under all the test conditions applied, a high temperature hydrocarbon fluid exercised a considerable advantage over other fluids with respect to time of ignition. The advantage was lost in tests where gasoline was added as a fire starter to the hydrocarbon oil. Tests to determine the ignitability of fluids are difficult to standardize and have been abandoned because of it. However, the ignition of a combustible material is the first step in any fire scenario and therefore is important to fire prevention.

Since arcing is the most common failure mode of devices incorporating insulating fluids, R. Hemstreet (1978) made an attempt to assess spray flammability. His experimental arrangement starts a flow through a nozzle and the liquid temperature is gradually increased until the spray sustains ignition at a measured distance of the igniter flame from the nozzle. The burning rate in a gas is proportional to the square of the gas pressure. Therefore, for two fluids of different viscosity, or the same fluid at different temperatures, the pressure in the aerosol cloud will be greater for the thinner fluid because more of it can be pumped through the nozzle in a given time. One would therefore expect that the minimum distance of the igniter flame to cause combustion would be larger in the case of a non-viscous fluid than in a viscous one or, equivalently, that the minimum igniter distance will increase for a given fluid as the temperature increases. This is shown to be the case in Hemstreet's results. Clearly, the experimental factors in an apparently simple test to rank fluids can be extremely complicated. For example, it was found that an air velocity of greater than about two feet per second had a considerable effect on the fluid temperature required for ignition as the nozzle distance was increased.

The overall shape of a temperature – distance curve in experiments of this sort are likely to be strongly influenced by the presence of decomposition products, either originally present in the fluid or induced by the igniter flame. The minimum temperature at which a liquid ignites depends on such factors as the degradation produced in the fluid by the flame and prior thermal and electrical stresses.

Heat release rate was measured in Hemstreet's work after removal of the ignition source are sometimes significantly different from results obtained using a sustained ignition source. Heat release rate data derived from measurements done on a quiescent pool of fluid provide information which have been used to calculate suggested clearances between the burning pool, presumably around the transformer, and the walls and ceilings of the room in which the unit is housed. The test does not consider the interference of a silica crust produced over a quiescent pool of silicone oil. In the unlikely event that the conditions necessary to maintain a quiescent condition were to exist in a real building fire, the wicking action of the crust causes the silicone oil to burn longer than high temperature hydrocarbon oils under the same conditions (Webber, 1983).

Despite the extensive work which has been done over the last few years on the validation of different fire hazard assessment methods for askarel replacement fluids there is still no consensus of opinion. Experience derived from the use of less flammable fluids, however, points to the fact that both hydrocarbon and silicone oils do perform satisfactorily and are suitable as askarel replacement fluids.

TRANSFORMER RETROFILL OPERATIONS

The retrofill operation, as it is conventionally practiced today consists of multiple stages, each of which involves draining, flushing and filling. The process can apply to two different situations, viz. (1) contaminated mineral oil transformers with >500 ppm PCB and (2) askarel filled transformers in which the coils are impregnated with PCB.

(1) In the case of a PCB contaminated mineral oil transformer it would be possible to reduce the PCB concentration of the working fluid by first draining the transformer and refilling it with non-contaminated oil. Since the PCBs are contained predominantly in the oil, and are not held by the porous insulation it is possible to retrofill the transformer to a required concentration level by repeated washings, as necessary. The effectiveness of the washings becomes increasingly important as the concentration of the original contamination increases since repeated dilution of residual contamination produces several times the original volume of the transformer tank for disposal. In addition, the presence of trapped, high concentration oil in the core/coil assembly, which is not easily removed by conventional flushing procedures, causes the bulk oil to become re-contaminated.

(2) In the case of an askarel transformer which is impregnated with PCBs, a simple flushing or vapor cleaning process leaves about 2.5% of the original askarel contained in the core/coil assembly. For a typical 250 gal. askarel transformer the impregnated fluid amounts to about 6 gal. A PCB concentration of 500 ppm PCB in a 250 gal. transformer is equivalent to only about one coffee cup in volume.

The equilibrium distribution of PCBs between transformer solid materials and the liquid dielectric in a 500 KVA transformer has been found to be about 97.5% in the bulk oil and 2.5% distributed unevenly among the paper, core steel and wire. The vast majority of the PCBs in the solid materials of an impregnated transformer are contained in the paper.

The diffusion of PCBs under the action of a strong concentration gradient from the interior of the paper towards the very low concentration in the bulk retrofill fluid is a slow process governed by Fick's laws. If the retrofilled transformer is not put into operation and remains at ambient temperature, then the viscosity of the askarel may be high enough to prevent diffusion at a significant rate. However, the EPA requires that the reclassification procedure should put the transformer under a normal load or, alternatively, that the bulk oil is heated to more than 50°C for a period of 90 days following the completion of the retrofill operation(s). The power losses of the transformer under load conditions produce heat which, in turn, reduces the viscosity of the impregnated askarel and increases its diffusivity.

If the concentration level of PCBs can be reduced to such an extent that subsequent leaching does not yield a concentration which is above the 500 ppm PCB limit imposed by the EPA 90 days after the process has been completed then, according to the letter of the law, the unit can be reclassified as 'PCB contaminated'. However, experimental data on units which have been retrofilled with a simple vapor cleaning process have shown that the PCB leaching rate under normal load conditions is in the range of 1.5-2.5 ppm PCB per day. If the trapped PCBs are able to leach out into the bulk retrofill fluid at a rate of as little as 1 ppm PCB per day, it would need approximately one year before the fluid concentration had climbed above 500 ppm PCB and the unit would have to be regarded once again as a PCB transformer.

The following discussion concerns the quantitation of the effect of repeated retrofill operations using a vapor cleaning system and describes how an alternative procedure can mitigate the leaching problem in a single operation.

The decision to retrofill rather than replace a contaminated transformer depends, as discussed above, upon a number of different factors. For example, the cost of new oil and the cost of labor have to be weighed against the benefits of reduced maintainance, service and disposal costs, the impact of public perception, and reductions in the possible costs of leaks, spills or fires. Once a decision has been made to retrofill, the usefulness of the action is governed by the effectiveness of the process to mitigate the problem of PCB leaching from the impregnated coil.

A cleaning process for the decontamination of PCB contaminated mineral oil in electrical equipment has been described in the form of guidelines by the Ontario Ministry of the Environment (1978). The stages involved are as follows:

* The transformer is emptied of bulk PCB by opening the drain valve. The unit is allowed to drain into an appropriate container for at least 24h.

* The drain valve is then closed or the unit sealed and filled with solvent and left filled for at least 24h.

* The solvent should then be drained and the transformer rinsed two more times by filling and draining with clean solvent.

* The second and third rinses may be retained for use as the first rinse in decontaminating other equipment.

The PCB solute in the mineral oil is distributed throughout the liquid and solid phases of the transformer. The PCB concentration remaining in the oil at equilibrium can be calculated from the partition coefficients for PCB between the transformer solid materials and the oil and the residual amounts of flushing fluid remaining in the tank between each stage. The effectiveness of this type of process is discussed more fully in a later section.

An alternative method of cleaning involves vapor degreasing the unit by spraying the cold internal surfaces with trichloroethylene vapor generated by boiling the solvent in an external vessel. The principle of operation is not unlike the soxhlet extraction of materials.

The process offers numerous advantages:

- the volume of contaminated solvent is kept to a minimum:
- the solubility of PCBs in the hot solvent is greater than in cold liquid and therefore the extraction process is less time consuming:
- the low viscosity solvent is able to penetrate the capillaries of the paper and core laminations and remove the PCBs from all the surfaces within the transformer.

Considerable work has been done on the solvent decontamination of electrical equipment and results have been reported by S.H. Hawthorne (1981). It was found that draining alone removed 91-96% of the PCB while a combination of draining and cleaning with trichloroethylene reduced the PCB concentration by 99.76%. Soaking the drained transformer with hot solvent, followed by vapor phase degreasing enabled 99.10% of the PCB to be removed. When an extended soak of 36 days was given to the transformer after the cleaning procedure had been completed the amount of PCB removed increased from the above figures to 99.94%, 99.96% and 99.72% respectively ie., the remaining PCB concentrations were 600 ppm, 400 ppm and 2,800 ppm.

Hawthorne's data allow some important conclusions:

* A simple transformer flushing procedure is effective in reducing the initial PCB concentration of an askarel filled transformer to less than 500 ppm.

* The PCB remaining after each cleaning process resided almost exclusively in the interstices between the steel core plates and in the paper and insulation of the copper coils, with little or no PCB remaining on the internal surfaces of the case and radiators.

* After the cleaning procedure had been completed the transformers were soaked for 36 days with the solvent from the third soak to remove the most rapidly leached PCBs. During this time it was found that the PCB concentration levelled out to a pseudo-plateau at 3500 ppm PCB in trichloroethylene after 10 days. In other words, 1000 ppm PCB had been leached out of the core/coil assembly in 10 days (100 ppm PCB/day). Also, the PCB concentration did not increase more than about 140 ppm PCB in the 26 days following the start of the plateau. This means that the leaching rate during this period is about 5 ppm PCB/day.

* The use of hot solvent was found to be more effective than cold solvent. Thus, when the retrofill had been completed and the transformer is put back under load, the heated oil can be expected to leach PCBs at a greater rate than if the transformer were not in operation. According to Hawthorne's results the initial leaching rate should be at least 5 ppm PCB/day. This data is corroborated in Figure (1).

* The lowest PCB concentration attained before the transformer was put back into service was 400 ppm PCB. If the leaching rate remained steady at 5 ppm PCB/day for the 90 days required by EPA, then the concentration would climb above 500 ppm PCB and the unit would remain a PCB transformer by definition. This assumes that the leaching rate is 5 ppm PCB/day. It is also assumed that the leaching rate remains constant at its initial, and therefore probably highest, rate. Even so, the results illustrate that a simple solvent flushing technique is able to reduce the PCB concentration into the range where the transformer, in the absence of leaching from the core/coil, could be re-classified as a PCB contaminated unit.

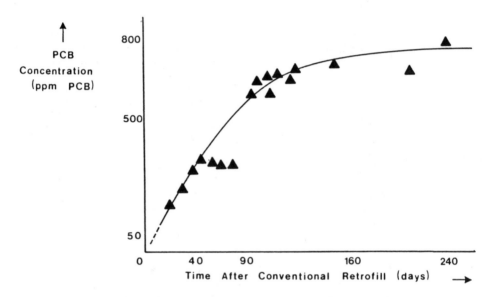

Figure 1. Conventional Freon Retrofill Technology Applied to
an Askarei Transformer.

Figure (1) shows a curve relating the increase in residual PCB concentration with time of transformer operation after the application of a vapor phase cleaning apparatus to a 750 kVA network transformer containing about 270 gal. of askarel. The curve consists of two main regions, an initially steep slope leading to a pseudo-plateau region which extends for more than 250 days. The initial slope of the curve is about 6 ppm PCB/day for the first 90 days after the transformer was put back in operation and about 1.3 ppm PCB/day thereafter.

The action of the vapour phase degreasing process is to initially dissolve the PCB on surfaces. Beyond this, capillary penetration takes place and the PCB in the outer layer of paper is replaced with solvent. When the transformer is filled with oil and put back into operation, the non-viscous solvent held in the paper quickly diffuses into the much larger quantity of more viscous dielectric fluid. During this time, the PCB concentration of the absorbed solvent is high because the concentration gradient which it sees is between an initial value of zero and 700,000 ppm PCB in the interior of the paper. The initial leaching of absorbed solvent into the oil therefore gives a rapid rise in bulk oil PCB concentration.

When the outer layer of trapped solvent has been replaced by slightly contaminated bulk oil there is again an almost constant concentration gradient but now the PCBs must diffuse through a medium which is about 450 times more viscous than the originally trapped solvent. Diffusion rate is inversely proportional to viscosity. Overall, the observed leach rate of PCB leaching will depend upon the relative viscosities of the solvent and oil (and hence temperature) and the depth of penetration of the solvent ie., the distance the PCB must diffuse.

The depth of penetration attainable by vapor degreasing alone is unknown. The results which have so far been obtained indicate that the leaching rate in the first 90 days is sufficient to increase the PCB concentration to more than the allowed 50 ppm PCB limit and, depending upon the efficiency of the operation, can be greater than 500 ppm PCB after 90 days. Thus, the depth of penetration of the degreasing solvent, using existing techniques, can only be very small and the main action of the process is to remove bulk PCB from tank walls and core/coil surfaces.

A MATHEMATICAL APPROXIMATION OF THE PCB LEACHING PHENOMENON

In an isotropic medium, Fick's first law of diffusion states that the amount of a substance diffusing through a medium is proportional to the concentration gradient acting as the driving force and a constant for the medium which is the diffusivity,

$$\text{i.e.,} \qquad X = -D \frac{\partial C}{\partial x} \qquad \text{............(1)}$$

where, X = mass of diffusing substance,
D = diffusivity,
C = concentration of diffusing substance,
x = distance.

Fick's second law of diffusion describes the flux of diffusing substance:

$$\frac{\partial C}{\partial t} = \frac{\partial}{\partial x}\left[D\frac{\partial C}{\partial x}\right]$$

where $\dfrac{\partial C}{\partial x}$ = concentration change measured in the x-direction

For the sake of approximation and mathematical simplicity assume that D, the diffusion coefficient, is constant.

$$\text{Then,} \qquad \frac{\partial C}{\partial t} = D\frac{\partial^2 C}{\partial x^2} \qquad \dots\dots\dots\dots\dots(2)$$

The diffusion of PCBs through insulating paper into the bulk oil of the transformer can be modelled as follows.

Assume that the x-y dimensions of a sheet of impregnated paper are large compared with its thickness so that edge effects can be ignored. Assume that the paper is homogeneous and of thickness 2d, and also that the bulk oil is maintained homogeneous.

The space occupied by the paper is $-d \le x \le +d$

The space occupied by the oil is $-d-a \le x \le -d$

and $d \le x \le +a$

where 'a' is the depth of the bulk oil in the x-direction.

The boundary conditions are:

$$t = 0, \qquad C = 0 \qquad \text{at } -d < x < +d$$
$$C = 0 \qquad \text{otherwise.}$$

At equilibrium, the rate at which PCBs leave the paper is equal to the rate at which PCBs enter the paper,

$$\text{i.e.,} \qquad a\frac{\partial C}{\partial t} = \pm D\frac{\partial^2 C}{\partial x^2} \qquad , x = \pm d, \ t>0 \quad \dots\dots(3)$$

The solution to a problem of this sort is given by J. Crank (1956) in terms of an infinite series as follows:

$$X_t/X_\infty = (1 + \alpha)*[1 - \alpha/(\pi^{0.5}*\beta^{0.5}) + \alpha^3/(2*\pi^{0.5}*\beta^{1.5}) - \alpha^5/(4*\pi^{0.5}*\beta^{2.5}) + \dots\dots] \qquad \dots\dots\dots(4)$$

where X_t = mass desorbed at time, t

X_∞ = mass desorbed at infinite time

$$\alpha = \frac{1 - X_\infty/X_S}{X_\infty/X_S}$$

where (X_S) is the total solute in the system.

(X_S) and (X_∞) are experimental quantities and therefore (α) can be obtained directly. β is a function of time and diffusivity defined as:

$$\beta = D*t/d^2.$$

Knowing (∞), a series of curves can be drawn relating (X_t/X_∞) with $(\beta^{0.5})$.

The experimental value of (∞) is matched to the corresponding theoretical curve of X_t/X_∞ vs $\beta^{0.5}$ and since $\beta = D.t/d^2$ a value can be obtained for the diffusion constant, D.

Measurements of the PCB content of impregnated dielectric papers show that a typical paper will contain $58 \pm 3\%$ by weight of an askarel such as Inerteen 70/30. Then, to calculate the total quantity of askarel absorbed in the paper one needs to approximate the amount of paper in a typical design.

For a 500 KVA transformer, the approximate quantities of the different types of materials are as follows:

Paper	$\sim$ 27 lb.
Core steel	$\sim$ 947 in.^{2}
Oil	$\sim$ 85 gal.

Thus, the amount of askarel impregnating the paper

$$= 12.3 \times 10^3 \times 0.58 \text{ g}$$
$$= 7,134 \text{ g}$$

If it is assumed that the pseudo-plateau shown in Figure (1) does allow a reasonable approximation of the total PCB leached into the bulk oil within the service lifetime of the transformer then,

$$X_\infty \sim \{X_s - [785 \text{ (ppm PCB)} \times 10^{-6} \times (0.878 \times 8.34) \text{ (lb./gal.)} \\ \times 270 \text{ (gal.)} \times 454 \text{ (g/lb)}]\}$$

If the paper quantity is assumed, for the purpose of calculation, to be proportional to fluid volume from one design to another then,

$$X_s = 12.3 \times 10^3 \times 0.58 \times 270/84$$
$$= 22,931 \text{ g}$$

therefore, $X_\infty = 22,931 - 705$
$$= 22,226 \text{ g}$$

i.e., $X_\infty/X_s = 0.969$

The calculation puts an upper bound on the extent of the leaching problem since the model used in the calculation assumes that sufficient time has been allowed to achieve equilibrium. In the experimental arrangement used, the time to reach equilibrium was not a limiting factor since this was long compared with the time required for the diffusion of PCBs into and out of single, as opposed to bundled, sheets of paper. A transformer, however, represents a different situation: the time to achieve true equilibrium may be beyond the service life-expectancy of the transformer since the migration of PCBs through different thicknesses of layered paper is the rate determining factor which governs the total fractional uptake.

The data shown in Figure (1) for a retrofilled askarel transformer suggest that PCBs will leach at the rate of about 1.3 ppm PCB/day for a period of perhaps years. It is therefore not possible to determine with any accuracy what mass of PCBs migrate into the oil when a final equilibrium is reached. If one assumes that, in the <u>best</u> case, a saturation level has been reached after 240 days then an approximate 'best case' correlation curve can be calculated between $\beta^{0.5}$ and X_t/X_∞ as follows.

From above, $\quad X_\infty/X_s \quad = \quad 0.969$

$$\alpha \quad = \quad \frac{1 - X_\infty/X_s}{X_\infty/X_s}$$

$$= \quad 3.2 \times 10^{-2}$$

$$\beta \quad = \quad \frac{D.t}{d^2} \qquad \text{by definition.}$$

Assume that $\quad D \quad = \quad 7.6 \times 10^{-10} \qquad (cm^2/sec)$
$\qquad\qquad\qquad\quad d \quad = \quad 1 \qquad\qquad\qquad (cm)$
then, $\qquad\qquad\quad \beta \quad = \quad 7.6 \times 10^{-10}.t \qquad (sec)$

If the transformer is to be reclassified as 'PCB Contaminated' then it must not be greater than 500 ppm PCB after 90 days of operation,

i.e., $\qquad \dfrac{X_t}{X_\infty} \quad = \quad \dfrac{500}{785} \quad = \quad 0.637$

By interpolation,

$$\beta^{0.5} \quad = \quad 0.036 \quad \text{when} \quad X_t/X_\infty \quad = \quad 0.638$$

i.e., $\qquad 0.036^2 \quad = \quad 7.6 \times 10^{-10}.t$
hence, $\qquad\quad t \quad = \quad 19.7 \text{ days}$

In other words, an initial clean layer of 1 cm of paper will allow the bulk oil to reach 500 ppm PCB in about 20 days.

The correlation shown in Figure (2) has been calculated between the depth of the cleaned layer and the time required for the PCB concentration of the bulk oil to reach 500 ppm PCB using different values for the diffusion coefficient.

The experimentally determined diffusion coefficient for the migration of Inerteen 70/30 through kraft paper is about (7.6×10^{-10}) cm^2/sec. Under these conditions, if the cleaned layer is about 4.3 cm deep, the transformer will become a PCB transformer one year after the retrofill has been completed.

If the transformer is to be reclassified as 'non-PCB' then it must not be greater than 50 ppm PCB 90 days after the retrofill procedure has been completed,

i.e., $\qquad \dfrac{X_t}{X_\infty} \quad = \quad \dfrac{50}{785} \quad = \quad 0.064$

hence, $\qquad\qquad t \quad \sim \quad 7 \text{ days}$

The correlation between the depth of the cleaned layer and the time to reach 50 ppm PCB is shown in Figure (3) for different values of the diffusion coefficient. Clearly, the time to reach a bulk oil concentration of 50 ppm PCB is much shorter for a given depth of cleaned

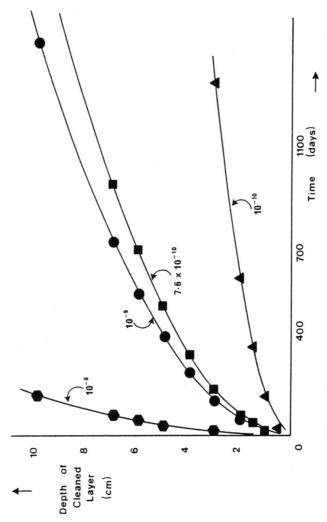

Figure 2. Recontamination of bulk oil to 500 ppm PCB.

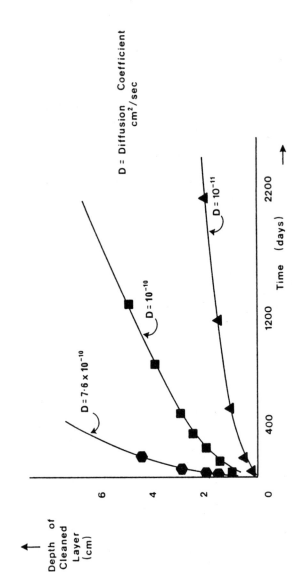

Figure 3. Recontamination of Bulk Oil to 50 ppm PCB.

layer and is critically dependent upon the diffusion coefficient.

Figure (4) shows how the depth of the cleaned layer must vary for different values of the diffusion coefficient in order to obtain a defined reclassification status. If the diffusion coefficient becomes larger than about 10^{-8} or 10^{-9} cm²/sec then the correlation shows that the depth of the cleaned layer must become much greater than it is practicable to achieve.

It is interesting to note that if the cleaned layer is 4.3 cm and the diffusion coefficient is the experimentally determined value of 7.6 x 10^{-10} cm²/sec. then there is little doubt that the transformer would be reclassified after 90 days. However, after one year of operation it would become a PCB transformer again by EPA definition _even if it had been reclassified as 'non-PCB'._

From the experimental curve of Figure (1), at t = 90 days,

$$\frac{X_{90}}{X_\infty} = \frac{560}{785} = 0.713$$

hence,

$$\beta^{0.5} = 0.0617$$
$$D = 7.6 \times 10^{-10} \text{ cm}^2/\text{sec (by experiment)}$$

hence,

$$d = 1.25 \text{ cm}$$

If the transformer is left in operation long enough to allow the viscosity of the residual impregnating fluid to increase due to the preferential migration of chlorobenzenes then, eventually, the diffusion coefficient will shift from

$$D = 7.6 \times 10^{-10} \text{ cm}^2/\text{sec} \underline{\quad\quad} D_\infty \sim 10^{-10} \text{ cm}^2/\text{sec}$$

The transformer will need to be drained and the oil decontaminated to less than 2 ppm PCB (EPA requirement) when these new conditions are met. An alternative would be to dispose of the contaminated oil as PCB fluid and use virgin oil to refill the transformer. There is considerable doubt that the diffusion coefficient will become low enough to merit this technique. A further drawback to the method lies in the fact that, during the long waiting period which would be necessary to observe a reduction in the rate of diffusion, the transformer would have to be considered PCB filled as though no retrofill operation had taken place.

The continued leaching of PCBs with a lowered diffusion coefficient of $\sim 10^{-10}$ cm²/sec. will cause the PCB concentration of the bulk oil to increase above 500 ppm PCB in a period of time which can be calculated as follows.

$$\frac{X_t}{X_\infty} = \frac{500}{785} = 0.637$$

hence,

$$\beta^{0.5} = 0.036 \quad \text{as found earlier}$$

therefore,

$$0.036^2 = \frac{10^{-10} \cdot t}{1.25^2}$$

i.e.,

$$t = 234 \text{ days}$$

Thus, even if the transformer is reclassified at the end of the EPA mandated 90 day period by a process which involves simple vapor cleaning, leaching and re-retrofill the technique cannot provide continued, reclassified use of the transformer for its remaining useful

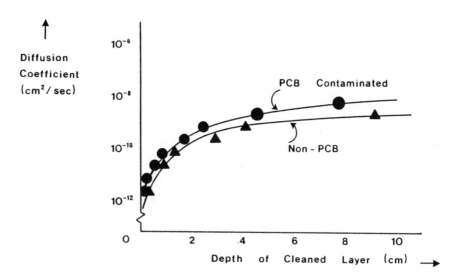

Figure 4. Depth of Cleaned Layer to Obtain Reclassification.

life. Indeed, it is not likely to be able to maintain the unit's reclassified status as a PCB contaminated transformer for as long as one year.

If it is hoped that the transformer can be reclassified as a non-PCB unit then,

$$\frac{X_t}{X_\infty} = \frac{50}{785} = 0.0637$$

then,

$$\beta^{0.5} = 0.0215$$

and

$$t = 84 \text{ days}$$

In other words, even if the transformer can be reclassified as non-PCB at the end of 90 days operation, it is likely that it will have to be considered as PCB contaminated very soon afterwards and that it will become a PCB transformer again in less than one year.

The relative importance of the adsorption of PCBs to their absorption has been assessed as follows.

The Freundlich adsorption isotherm is represented by the equation,

$$\frac{X}{M} = k \; C^n \qquad \ldots\ldots\ldots\ldots(5)$$

where X and M are the masses of adsorbate and adsorbent, respectively and k and n are Freundlich coefficients. C is the equilibrium value of the PCB in the bulk oil.

Experiments were conducted at 50^0C since this is the temperature required by the EPA when a retrofilled transformer is to be reclassified and also at 110^0C representative of the top oil temperature. The following table shows results obtained for two different types of dielectric papers.

Table (1)

Freundlich Adsorption Coefficients

Paper	Temperature (^{0}C)	Freundlich Coefficients k	n
Dicy	50	0.49	1.05
	110	2.75	0.77
Diamond	50	1.32	0.78
	110	0.04	1.22

$$X/M = 1.32. \; C^{0.784} \quad \text{for diamond paper at } 50^0C$$

$$X \sim 200 \text{ g}$$

Thus, the quantity of PCBs adsorbed on the paper is extremely small relative to the total quantity of PCBs absorbed by the paper.

Figure (5) illustrates the effect which processing can have on

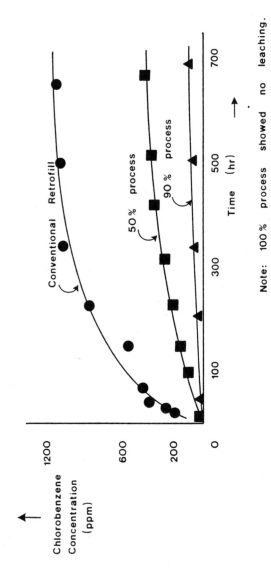

Figure 5. Trichlorobenzene Leaching as a Less Viscous
Surrogate for PCBs.

Note: 100 % process showed no leaching.

mitigating the leaching problem. The data were obtained using chlorobenzenes as a non-viscous surrogate for PCBs in order to more easily quantitate any observed leaching. The completed process allowed permanent reclassification of the bulk retrofill fluid.

The model used virgin transformer materials impregnated with chlorobenzenes. The bulk oil was homogenized before samples were taken and the temperature of the tank adjusted to 50°C. Samples of impregnated kraft paper were wound into a coil and immersed in the oil. Several different kinetic runs were made with different thicknesses of paper as well as different numbers of layers forming the overall wad of paper.

The experimental conditions shown in Figure (2) were found convenient because a series of complete leaching curves could be compared over a period of about 30 days rather than the 300 days found for a 750 KVA transformer. It is clear that sufficient time has elapsed to be able to compare a conventional vapor cleaning retrofill system with the processing necessary to avoid the leaching problem.

It was found that processing beyond simple vapor cleaning could be continued to the point where leaching was completely stopped. In actual practice, however, such a rigorous level need not be achieved since, if the leaching rate is reduced by, say, an order of magnitude from the measured value of 1.3 ppm PCB/day then, instead of about one year of reclassified use, the unit would retain its classification for 10 years. In this example, if the transformer's life-expectancy was only for a further 5 years then clearly it would remain reclassified for the remainder of its useful life. The extent of processing can be tailored to the level of reclassification required and the life-expectancy of the transformer.

A method which is often used for the PCB decontamination of mineral oil transformers is to drain the fluid into drums for disposal, flush and drain the transformer with a flushing oil and then refill the unit with uncontaminated mineral oil. The volume of flushing oil needed to reduce the PCB concentration of the retrofilled transformer to less than 50 ppm PCB depends upon the initial concentration of the oil and the amount of residual oil left in each flushing operation. The PCB concentration after a drain and flush sequence can be predicted according to the following equation:

$$C_1 = C_0*(R/100 + P_s/P_1)*(1 - P_s/100)$$

where

C_1 = PCB concentration in oil after draining and refilling.

C_0 = PCB concentration in oil before draining and refilling.

R = Percent of fluid remaining in the transformer after draining.

P_s = Percent of PCB in solid materials before draining.

P_1 = Percent of PCB in oil before draining.

The mathematics assumes that the oil in the tank is homogeneous. In a transformer, however, the volume of oil trapped in the interstices of the windings are not likely to be removed by a simple drain and flush procedure. Then, when the unit is put back into service, the trapped, highly PCB contaminated original oil gradually convects into the retrofilled bulk oil to give an apparent leaching effect. If the trapped, original oil had a PCB concentration of 2000 ppm PCB and becomes diluted by a factor of 40 by convection then the final concentration of the retrofilled fluid will be 50 ppm PCB.

The higher the PCB concentration in the mineral oil the more likely it will become necessary to remove trapped oil in order to achieve a retrofilled concentration of less than 50 ppm PCB. The areas which must be cleaned in particular are the core/coil assembly and other regions which contain flow restrictions.

It has been mentioned above that a vapor cleaning system has been proven successful in removing surface PCB contamination from areas previously considered inaccessable. The method has the advantage that a minimum quantity of PCB contaminated fluid is derived for disposal in contrast to the drain and flush method which produces several times the volume of the transformer tank. Also, as the level of PCB contamination in the mineral oil increases the vapor cleaning method becomes increasingly necessary and eventually becomes the only viable cost-effective procedure.

Chemical reagents which are capable of reacting specifically with an unreactive component of the oil, namely PCBs, while leaving other essential components intact, must be carefully researched in terms of chemical reactivity, safe handling, storage capability, toxicity and cost. The following discussion gives a brief overview of some of the work which has so far been completed in the development of the Mini-Oil-Plant (MOP) system for the PCB decontamination of mineral oils.

Chemical reaction systems which operate by a free radical mechanism typically produce a sludge which must be separated from the oil before the fluid can be reclaimed to dielectric quality (Webber et al. 1984).

The fact that a sludge is produced in chemical reaction systems may indicate that, in addition to a reaction taking place between the reagent and PCBs, there may also be reactions with components of the oil. Some of the components which are removed may be beneficial to the performance of the oil as an insulating dielectric. This is discussed in detail below but it can be briefly stated here that the re-usability of the oil from the MOP process can be substantiated in both experimental and theoretical terms.

For the most part, an insulating oil contains a complicated mixture of chemically inert alkanes. The most immediate reaction between the reagent and the components of an unoxidized oil is that both naturally occurring and added oxidation inhibitors will be removed.

The mode of action of an oxidation inhibitor is to react with an oxygenated, reactive free radical in solution to produce a free radical derived from the oxidation inhibitor. For a given compound to inhibit the oxidation process its free radical must be stable relative to the oil free radical from which it was formed. This is found to be the case with compounds such as di-t-butyl-p-cresol (DBPC) or di-t-butyl phenol (DBP) (Webber and Wilson, 1981).

The chemical phenomena which are known to be responsible for the effectiveness of a given compound towards oxidation inhibition (e.g., delocalization of the free radical electron) indicate that aromatic phenols are likely to be effective. The total aromatic content of a typical insulating oil is 10-15% but only a very small part of this would be expected to be phenolic. The majority of oxidation inhibitor in a dielectric quality oil is expected to be not more than 0.3% in accordance with IEEE specifications for new oil. Hence, the sludge formed in a chemical reagent process is expected to contain excess reagent and oil insoluble compounds resulting from reaction with dissolved water, ketones, acids and oxidation inhibitors.

The elimination of organically bound chlorine from the PCB molecule as inorganic chloride is a reduction process which involves an initial transfer of an electron from the reagent to the PCB molecule. Once an electron has been transferred to the PCB, the aromatic nucleus will contain a single unpaired electron. This species is known as a 'radical anion'.

In general, radical anions are effective in abstracting a hydrogen atom only from those compounds which would typically be polar relative to the undegraded oil components (Webber, 1983). Consequently, chemical treatments which employ organometallic reagents tend to react with not only the PCBs present but also oxidation products and oxidation inhibitor. The result is that one would expect that an oxidized, PCB contaminated oil would become oxidation product free as well as PCB decontaminated. The problem, however, lies in the reactivity of the reagent towards PCBs. In the absence of a solvent the reaction is typically very slow at convenient operating temperatures. In the presence of solvent the cost of oil reclamation becomes too high because even slight traces of polar solvent will destroy the dielectric properties of the oil. The MOP process does not use a polar solvent and does not compromise oil quality.

An important feature of reclamation is whether or not reclaimed oil is equivalent to new oil and in particular, whether or not PCB decontaminated oil is equivalent to new oil.

It has been shown that fuller's earth treatment of partially oxidized oils is primarily effective in reducing the neutralization number and power factor to acceptable levels but is relatively less effective in producing the values of interfacial tension, color, dielectric breakdown, etc., typical of new oils (Webber, 1980).

Fuller's earth treatment does not remove PCBs or DBPC oxidation inhibitor. The PCB decontamination process, on the other hand, involves a highly reactive reagent which has a strong affinity for the most easily reduced components of the oil. The following discussion considers the chemical effects of organometallic reagents on the classes of compounds normally encountered in transformer oils.

Alkanes

Oils which are totally paraffinic are essentially inert to radical ions. For example, when a completely paraffinic oil is used to disperse a suspension of reagent, oil properties, such as power factor, remain unaffected. The power factor of the oil is a sensitive measure of its polar constituents so that if a reaction between the reagents and the oil components does take place it is important that they at least do not yield polar products. The extent of possible interactions has been investigated by various analytical techniques including capillary gas chromatography, mass spectroscopy, Fourier transform infra-red spectroscopy and Fourier transform nuclear magnetic resonance. No significant difference was observed between the alkane constituents of reacted and unreacted mineral oils in the MOP process.

Alkenes

This class of compounds is not expected to occur in significant

quantities in transformer oils although, if they were present, they
would most probably undergo their most common reaction of polymerization
and precipitate out of the oil into the sludge layer. An alternative
possibility is that the olefin may add to the alkyl side chain of an
alkyl aromatic compound. In either case it is most unlikely that the
products of such reactions would deleteriously affect oil properties and
indeed such reaction products may occur naturally in some oils.

Polychlorinated Biphenyls and Chlorobenzenes

The reaction of the reagent with halogenated compounds in a
contaminated insulating oil should, according to the mechanisms proposed
by Sargent (1971), Pilgrim and Webber (1981), and Wilson and Webber
(1981), produce polyphenyls, including biphenyl. Polyphenyls with four
or more phenyl groups are insoluble even in hot oil and are expected to
be the major constituents of the reaction sludge. This has been verified
by extensive capillary gas chromatographic/mass spectroscopic analyses.

Aldehydes and Ketones

Aldehydes are a class of compounds which are not expected to be
present in oxidized oil in significant quantities because of their
reactivity. Ketones, on the other hand, are known to be present as
primary oxidation products.

The oxidation of hydrocarbons produces, successively,
hydroperoxides, alcohols and ketones, and finally acids. The fuller's
earth reclamation process is effective in the removal of acids but is
relatively ineffective in removing their precursors. The presence of
even small quantities of acid can cause an autocatalytic effect in the
decomposition of ketones to give further acids. Oil which has been
reclaimed by conventional fuller's earth treatment is therefore likely
to oxidize more rapidly than new oil and this is usually found to be the
case.

The reactions of aromatic ketones with radical anions lead to
reduction of the ketone and yield an alcohol as the major product. For
example, benzophenone produces benzhydrol in 90% yield and p-
benzoquinone yields hydroquinone in 92% yield. The reactions of
aliphatic ketones, on the other hand, produce alcohol derivatives
resulting from addition to the anion.

It was described earlier that the interaction between radical
anions and ketones in a partially oxidized, PCB contaminated oil,
produces alcohols. It would therefore be expected that the formation of
aromatic alcohols in particular would give rise to 'natural' oxidation
inhibitors. It is very likely that the effectiveness of the inhibitors
formed in this way is small, relative to DBPC but, in any case, the oil
should not be deleteriously affected by the treatment provided that the
reclaimed oil has DBPC added to it. The oxidation stability of the
processed oil should then be as good as, or possibly better than, new
oil. Experimental data have shown that the oxidation stability of oils
treated by the MOP system are the same as new oil by ASTM D2112.

THE DISPOSAL OF ASKAREL FLUIDS

The methods in use for the destruction of fluids which contain high
(percentage) levels of PCBs are typically large energy sources. The most
widely used method for the disposal of askarel fluids is high
temperature incineration. Other methods include reaction with molten

sodium carbonate, ozonolysis, and chlorolysis.

In contrast with these methods is a chemical reaction which has been found to completely dechlorinate transformer askarel in 4 minutes at room temperature. The proprietary reagent, packed into a column configuration, allowed complete dechlorination of aroclor 1260 during the 1 min. residence time of the reacting PCB. The process is being investigated to allow the treatment of askarel from retrofilled transformers to be done on-site with a mobile system. The advantage which this process offers is that, firstly, transformers can be retrofilled and reclassified in a single operation and, secondly, a certificate of destruction for the askarel can be supplied to the customer before the service crew has left the job site. The customer then does not have the liability of the transportation and interim storage of PCBs at a waste disposal facility.

CONCLUSION

Diffusion coefficients have been measured and used to calculate the rate of PCB leaching from impregnated insulation into oil used to retrofill askarel transformers. It has been shown that it is theoretically unlikely that a simple vapor cleaning system alone can be used to reclassify a transformer to 'non-PCB' status.

It is particularly pertinent to note that even if a reclassification is achieved, it is improbable that the status will be retained for longer than one year before the unit must again be considered to be a PCB transformer. In addition, if the transformer is re-retrofilled, the calculations show that it is still unlikely that the unit will retain a reclassified status for a significantly longer period of time.

Transformers which contain contaminated mineral oil with greater than 500 ppm PCB are defined as PCB transformers. This type of unit can be successfully reclassified as 'non-PCB'. The use of the vapor cleaning method has been contrasted with the 'conventional' method of retrofilling this type of transformer using a drain and flush procedure. The use of a vapor cleaning system enables the removal of pockets of the original highly contaminated oil trapped in interstices and the method thereby avoids the recontamination of the retrofill oil by convective mixing.

The vapor cleaning method becomes a practical requirement when the concentration of the PCB contamination is greater than about 1000 ppm PCB. The method has a significant cost advantage over the conventional drain and flush process since vapor cleaning yields a minimum volume of contaminated fluid for disposal.

REFERENCES

Albro, P.W., Hass, J.R., Crummett, W.B. 1979. Ann. N.Y. Acad. Sci. 320, 125: Summary of the workshop on recent advances in analytical techniques for halogenated aromatic compounds.

Allen, J., and Norback, D. 1973. Science 179, 498: PCB and PCT-induced gastric mucosal hyperplasia in primates.

Ballard, J.G. and Hawthorne, S.H.: 'Solvent Decontamination of PCB Electrical Equipment', Canadian Electrical Association Engineering and Operating Division: 1981 Part 1, 81-A-66.

Bertazzi, P., Zocchetti, C., Guercilena, S., Foglia, M., Pesatori, A., and Riboldi, L. 1981. Int. Symposium on Prevention of Occupational Cancer, Helsinki, Finland April 21-24, 1981. Mortality study of male and female workers exposed to PCBs.

Brown, C.E. and Webber, I. 1982. Unpublished results and also, Brown, C.E., Bezoari, M.D. and Kovacic, P. 1982. J. Polymer Science, 20, 1697: Characterization of aromatic polymers (benzenoid and heterocyclic) by cross-polarization, magic-angle C-13 NMR spectroscopy.

Buser, H.R. and Rappe, C. 1979. Chemosphere, 8, 157: Formation of PCDFs from the pyrolysis of individual PCB isomers.

Buser, H.R. 1979. Chemosphere, 8,415: Formation of PCDFs and PCDDs from the pyrolysis of chlorobenzenes.

Canadian Ministry of the Environment: Pollution Prevention and Waste Management Guidelines for Polychlorinated Biphenyls: November 1978.

Chen, P.H.-S., Luo, M.-L., Wong, C.-K., Chen, C.-J. 1984. Am. J. Ind. Med. 5, 133: PCBs, PCDFs and PCQs in Yu-Cheng oil and PCBs in the blood of Yu-Cheng patients.

Chittim, B.G., Clegg, B.S., Safe, S.H., and Hutzinger, O. 1979. Environment Canada report 05578-00067: PCDFs and PCDDs: Detection and quantitation in electrical equipment and their formation during the incineration of PCBs.

Crank, J.: 'The Mathematics of Diffusion': Clarendon Press, 1956.

Doble client Conference, Doble Corp. 1982: summary of replies to the Transformers section of the Technical Questionnaire.

EPRI report #325, 1976: Distribution Transformer Tank Pressure Study.

Goldstein, J. 1980. In Kimbrough, R.D., ed., Halogenated Biphenyls, Terphenyls, Naphthalenes, Dibenzodioxins and Related Products. Elsevier/North-Holland Biomedical Press, New York.

Hawthorne, S.H. and Ballard, J.G. 1981: Canadian Electrical Association Engineering and Operating Division, 'Solvent Decontamination of PCB Electrical Equipment' Part 1, 81-A-66.

Hemstreet, R.A. 1978: 'Flammability Tests of Askarel Replacement Transformer Fluids', FMRC report, Serial No. 1A7R3.RC prepared for National Electrical Manufacturers Association.

Hori, S., Obana, T., Kashimoto, T., Otake, T., Nishimura, H., Ikegami, N., Kunita, N., and Uda, H. 1982. Toxicology, 24, 123: Effect of PCBs and PCQs in cynomologus monkey (Macaca fascicularis).

Hsu, S.-T., Ma, C.-I., Hsu, S.K.-H., Wu, S.-S., Hsu, N.H.-M, and Yeh, C.-C. 1984. Am. J. Ind. Med. 5, 71: Discovery and epidemiology of PCB poisoning in Taiwan.

Humphrey, H. 1983. D'Itri, F.M., and Kamrin, M.A., eds., PCBs: Human and Environmental Hazards. Butterworth Publishers, Boston.

Hutzinger, O., Safe, S., and Zitko, V. 1974. CRC Press Cleveland, Ohio. The Chemistry of PCBs.

Kimbrough, R. 1981. In Khan, M.A.Q., and Stanton, R.H., eds., Toxicology of Halogenated Hydrocarbons: Health and Ecological Effects. Pergamon Press, New York.

Kunita, N., Kashimoto, T., Miyata, H., Fukushima, S., Hori, S. and Obana, H. 1984. Am. J. Ind. Med. **5**, 45µ Causal agents in Yusho oil.

Kuratsune, M. 1972. Environ. Health Perspect. **1**, 129: Laboratory examinations of patients with Yusho.

Lao, R.C., Thomas, R.S., Monkman, J.L. 1976. Dynamic Mass Spectrom. **4**, 107: Application of computerized GC–MS to the analysis of PCBs.

Masuda, Y. and Yoshimura, H. 1984. Am. J. Ind. Med. **5**, 31: PCBs and PCDFs in patients with Yusho and their toxicological significance.

Mescher, J.A., Carnes, R.A., Duvall, D.S. and Rubey, W.A. 1978. USEPA Research Brief: Thermal degradation of PCBs.

National Electrical Manufacturers Association, 1980: A report on research on fire safety test methods and performance criteria for transformers containing PCB replacement fluids. Contract no. NB795BCA0024.

Pilgrim, D. and Webber, I. 1980. Simon Fraser University work term report: PCB dechlorination treatments for insulating oils.

Poland, A., and Glover, E. 1977. Mol. Pharmacol. **13**, 924: PCB induction of aryl hydrocarbon hydroxylase activity: A study of the structure–activity relationship.

Rappe, C., Marklund, S., Berggvist, P., and Hansson, M. 1982. Chem. Scripta., **20**, 56: PCDDs, PCDFs and other polynuclear aromatics formed during PCB fires.

Rappe, C., Marklund, S., Berggvist, P., and Hansson, M. 1983. In Choudhary, G., Keith, L.H., and Rappe, C. eds., Chlorinated Dioxins and Dibenzofurans in the Total Environment. Butterworth publishers, Boston: Polychlorinated dibenzo-p-dioxins, dibenzofurans and other polynuclear aromatics formed during incineration and polychlorinated biphenyl fires.

Sergent, G.D., 1971: 'Proposed Mechanism for the Reductive Dechlorination of PCBs': Tetrahedron Letters, 3279 (1971).

Sherma, J. 1975. Advances in Chromatography, Giddings, J.C., Gruehka, E., Keller, R.A., Cages, J., eds. New York: Marcel Dekker, Inc. Chapter 5: GC analysis of PCBs and other nonpesticide organic pollutants.

Smith, R.M. et al. 1982. Chemosphere, **11**, 715: TCDF and TCDD in a soot sample from a transformer explosion in Binghamton, New York.

Stalling, D.L., Huckins, J.N., Petty, J.D., Johnson, J.J., Sanders, H.O. 1979. Ann. N.Y. Acad. Sci. **320**, 48: An expanded approach to the study and measurement of PCBs and selected planar halogenated aromatic environmental pollutants.

Unger, M., Olsen, J. and Clausen, J. 1982. Environ. Res. <u>29</u>, 371: Organochlorine compounds in the adipose tissue of deceased persons with and without cancer: a statistical survey of some potential confounders.

Vos, J.G., Koeman, J.H., Van der Maas, H.L., Ten Noever de Braun, M.C., and de Vos R.H. 1970. Fd. Cosmet. Toxicol. <u>8</u>, 625: Identification and toxicological evaluation of PCDDs and PCNs in two commercial PCBs.

Webber, I. and Wilson, D.E. 1981. University of Victoria work term report: Inhibition of PCB dechlorination by 2,6-di-tertiarybutyl-p-cresol.

Webber, I. 1983. CEA report 77-70 phase II: The radiolytic degradation of PCBs in electrical insulating oils.

Webber, I 1983: Assessment of fire hazard safety for askarel substitute fluids: RTE Research and Development in collaboration with Underwriters Laboratories.

Webber, I., Pilgrim, D. and Thompson, M.A. 1984: 'The Safe Disposal of PCBs': IEEE/IAS Transactions, Vol. IA-20, No.1 p. 159 Jan./Feb. (1984).

Webber, I., 1980: 'A Study of the Factors Affecting the Reclamation of Degraded Electrical Insulating Oil': Canadian Electrical Association Report 77-70, Phase I (1980).

Williams, C., Prescott, C., Stewart, P., and Choudhary, G. 1984. In Keith, L.H., ed. Chlorinated Dioxins and Dibenzofurans in the Total Environment. Butterworth Publishers, MA.: Formation of PCDFs and other potentially toxic chlorinated pyrolysis products in PCB fires.

DESTRUCTION

PCB DESTRUCTION

Jacques Guertin

Electric Power Research Institute
3412 Hillview Avenue
Palo Alto, CA   94304

Abstract. Polychlorinated biphenyl (PCB) can be destroyed in a variety of ways. Currently, for PCB concentrations >500 mg/kg, incineration at an ANNEX I (or equivalent) facility is generally the only EPA approved method. There are additional options for destruction of liquids containing PCB at concentrations <500 mg/kg. Use of a high-efficiency boiler is an attractive option. Also, EPA has selectively approved chemical dechlorination for PCB concentrations <1%.

PCB-contaminated soil must be cleaned up (to a site-specific concentration), usually by excavation and removal followed by incineration (ANNEX I facility). Other options such as soil washing (extraction) or in situ PCB destruction require demonstration prior to EPA approval.

INTRODUCTION

PCB has been implicated as a hazardous material. Existing regulations require various actions depending on the PCB concentration and its location (the exposure risk). Utilities must dispose of PCB capacitors exposed to the public by October 1988. (Capacitors in restricted access areas may be used for the remainder of their useful life). At all times, there is a risk of a PCB capacitor or PCB transformer fire (or spill) resulting in possible contamination of the environment with PCB or PCB-related compounds such as polychlorinated dibenzofuran (PCDF) or polychlorinated dibenzo-p-dioxin (PCDD). Consequently, many owners of PCB-containing equipment or materials wish to dispose of the PCB. This chapter examines the various options of PCB disposal, focusing on permanent removal, i.e., destruction.[1] Figure 1 shows the molecular structure of the compounds of environmental concern.

PCB DESTRUCTION: METHODS

Contained PCB

PCB stored in equipment, devices, and various products, is considered contained, i.e., it has not escaped into the

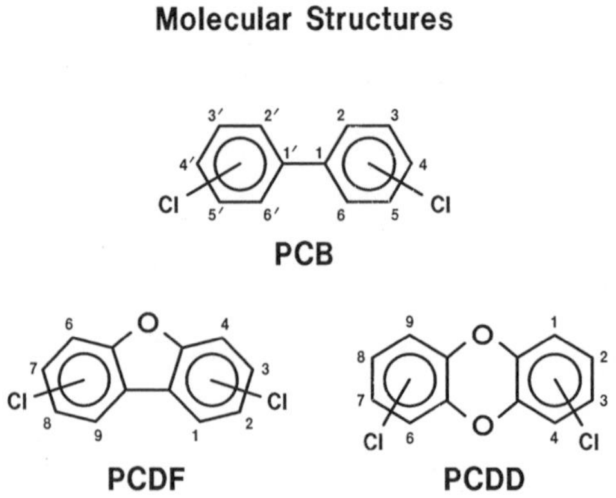

Molecular Structures

Fig. 1. Molecular structures of PCB and the PCB-related compounds, PCDF and PCDD. There are 209 different PCB, 135 different PCDF, and 75 different PCDD compounds (including the monochlorinated compounds in each case) depending on the specific position of the chlorine atom(s) in the molecule.

environment. (If it is necessary or desirable to destroy such PCB), two general approaches are considered: thermal and nonthermal. Disposal in a landfill does not of itself destroy PCB and is not permitted for PCB concentrations >500 mg/kg.

<u>Thermal Destruction</u>: Laboratory tests indicate that a temperature >1173 K over a residence time[*] >2 s will completely destroy PCB in the presence of excess oxygen.[2,3] For additional safety, regulations call for 1473 K or 1873 K and a residence time of 2 s or 1.5 s, respectively. These conditions are more than sufficient to destroy PCB-related species of concern such as PCDD and PCDF. The major products of combustion are CO_2, H_2O, HCl, and a small amount of Cl_2.

Currently, incineration (ANNEX I specification) is the most commonly chosen method to destroy PCB. In some cases (e.g., large PCB capacitors or PCB concentrations >500 mg/kg), it is the only EPA approved method (PCB manual[1]). Measurements have shown >99.9999% destruction efficiency. However, regulations are not based on efficiency. Rather, the maximum allowable stack emission of PCB is 1 mg per kg of PCB input. Solid residues (as a result of incineration) having PCB concentrations >50 mg/kg must either be reincinerated or placed in an EPA approved landfill. There are six commercial incinerators in the United States approved by EPA to burn PCB:

o   Rollins

o   Energy Systems Company (ENSCO)

o   Pyrotech Systems/ENSCO

o   Chemical Waste Management

o   General Electric

---

[*]Defined as volume of combustion chamber divided by volume flow rate of gas leaving chamber.

Rollins (Deer Park, TX) has an incineration facility permitted to burn PCB. Waste can be delivered to this site by rail, road vehicle, or barge. Also, Rollins operates a chemical waste landfill on the same site. Thus, combustion residue and decontaminated transformers can be buried there. The facility can incinerate all forms of PCB including PCB in "solid" form such as a capacitor. A PCB capacitor is first shredded and the pieces are incinerated at 1313 K in a rotary kiln. Volatiles are further incinerated in an after-burner operating at 1573 K. The afterburner residence time is about 2.5 s.

ENSCO (El Dorado, AK), like Rollins, combines a rotary kiln with liquid injection incineration. However, the shredder-kiln is totally enclosed. Thus, no PCB can escape during the shredding operation.

Pyrotech Systems/ENSCO (Tullahoma, TN) has a truck-mounted incinerator based in El Dorado, TX. Oxygen-rich combustion occurs at a temperature >2473 K with a residence time of 2 s. The facility accepts PCB liquid at concentrations <50%.

Chemical Waste Management (Oak Brook, IL) recently bought SCA Chemical Services, Inc. Their incinerator can accept liquid or soil waste (including capacitors). Generally, the chlorine content of total organic waste should be <20%.

General Electric's "thermal oxidizer" incinerator (Pittsfield, MA) accepts PCB-contaminated fluids. The average PCB destruction efficiency is 99.99998%. Stack gases are scrubbed such that 99.8% of the HCl is removed. No PCB is detected in the scrubber water (measurement detection limit is 1 ng/g).

Ocean incineration has the inherent advantage of reducing the danger of exposure to the general populace. Therefore, regulations are less stringent (e.g., no removal of HCl is required). However, there are also difficulties: (a) corrosion from sea salt, (b) high waves, (c) storage of solid ash residue, (d) stack height designed to minimize long-range transport of emissions is in conflict with that designed to protect the workers, and (e) workers must be protected from incinerator heat. Presently, Chemical Waste Management is negotiating for a permit to incinerate PCB using the ship, Vulcanus. The ship has a PCB destruction efficiency of 99.99%.

Destruction of PCB-contaminated fluid in a high-efficiency boiler (>99% combustion efficiency) is an alternative to ANNEX I incineration. EPA permits the use of such a boiler since it operates at lower cost and can help reduce the backlog of stored PCB. It is a particularly attractive approach for owners of PCB fluids on the same site as their boilers since no transport of PCB is needed and the thermal energy of the PCB molecule is effectively used. (This is partially offset by additional corrosion as a result of generated HCl.) EPA regulations indicate that the PCB fluid must be <10% (by volume) of the total fuel and not be added during startup or shutdown operations. The PCB-contaminated fluid is most often a mineral oil, the PCB concentration of which must be <500 mg/kg. (For fluids other than mineral oil PCB carriers, the monitoring requirements are more stringent.) The key to successful boiler destruction of PCB is adequate mixing. Boilers with tangential, opposite-wall fuel nozzels achieve excellent mixing. Nominal temperatures and residence times are 1703–1948 K and 1–3.7 s, respectively.

Very few utilities are currently burning PCB-contaminated oil in their boilers even though numerous tests have shown destruction efficiencies >99.9999% (even for inputs of PCB concentrations in mineral oil as high as 5%!). The problem is most commonly public concern regarding a perceived problem. And this is in spite of the fact that the typical input rate of wastes oil/fuel ratio is only 0.001, meaning that airborne emissions of PCB would be negligible even if no destruction occurred in the boiler combustion zone!

Using lower temperatures than conventional combustion (but longer residence times), molten-salt destruction has achieved 99.9999% PCB destruction efficiency. The salt is typically sodium carbonate (Na_2CO_3). Because emitted particles and vapors are absorbed or react with the salt, fewer emmission controls are necessary. The process destroys liquid or "solid" PCB; however, the salt bed may subsequently become contaminated by noncombustible residue.

Many of the principles and advantages of incineration using molten salt also apply to destruction using a molten-glass bed. For example, there are no flameout problems because temperature is maintained by electrical heating no matter what happens to the input material or whether the input material has a low energy content. Solids dissolve in the glass for ultimate safe disposal or use. Since the glass is less reactive than salt, chlorine compounds (such as HCl and $CaCl_2$) can be recovered and sold. Penberthy Electromelt Company (Seattle, WA) operates such a facility although it is not yet commercial.

In an Electric Power Research Institute (EPRI) project, a direct-current arc furnace has been constructed (Arc Technologies, Model City, NY) especially for destroying whole capacitors. The intact capacitor need only be punctured (no shredding) after being placed in the furnace to avoid the possibility of a pressure explosion. The arc furnace contains molten iron-aluminum at 1873 K and is designed such that all gases must exit through the arc. Inherent ultraviolet (UV) radiation also contributes to destruction of organics. The process is essentially a pyrolysis with all materials being uniformly exposed to high temperatures. Therefore, the necessity for "scrubbing" is much reduced because no air is used in the destruction process. A permit is being sought to perform test "burns".

Nonthermal Destruction: Three processes for nonthermal destruction of PCB have been applied: (a) chemical, (b) UV radiation, and (c) biological. These less-violent procedures have the potential advantage of stripping the chlorines from the PCB molecule leaving unchlorinated biphenyl and achieving a decontamination of originally PCB-contaminated oil. Thus, the oil can be reused. (Nonthermal methods are not very effective for PCB concentrations >>10,000 mg/kg.)

Chemical approaches for PCB destruction are generally based on reactions with some form of alkali metal (sodium[*] as

---

[*]Pure sodium is hazardous since it reacts violently with water; Na-naphthalide and Na-polyethylene glycol are examples of less-reactive species.

an organometallic, NaR, and sodium or potassium as a metallic alkoxide, NaOR or KOR[**]). The chlorine is removed as an inorganic salt. The following is a partial list of companies using this technology commercially: Chemical Clean, Acurex, Sunohio, and PPM, Inc. (A. D. Myers and Sanefen use a chemical process without sodium or potassium.) ERT (with Tracer Technologies) is planning to commercialize an electrochemical process.

Although UV and biological processes are known to decompose PCB to some extent,[4,5] there is only one commercial application. Detox Industries, Inc. (Sugar Land, TX) has received EPA permission (Region 6) to use a biological approach for degradation of PCB in soil and sludge. The process can be applied in situ or on excavated material and is claimed to produce a PCB half-life of two weeks even for PCB concentrations as high as 10%!

Finally, for mineral oils contaminated with PCB at concentrations <1000 mg/kg, it may be economical to simply remove the PCB from the oil and reuse the oil. EPRI has examined a solvent extraction approach. As a result of this study, a southeast utility has constructed a processing plant designed to handle 1900 m^3/y. The plant reduced PCB concentrations by a factor of 100 and the concentrated residue (approximately 5% PCB) is sent to an incineration facility.

PCB Spills

Although liquid PCB from a spill would not be expected to migrate large distances away from its source (strong absorption by the carbonaceous component of soils), it nevertheless must be removed from the soil and/or groundwater to prevent it from getting into the food chain. The standard procedure is to excavate the soil until the PCB concentration becomes the same as that of the background soil (for frequent animal or human contact areas) or to perhaps <50 mg/kg otherwise.[6]

[**]R represents any organic molecular group (metal-carbon or metal-oxygen-carbon molecular bonding).

On-site soil treatment is an alternative to excavation, removal, and incineration. The JM Huber Corporation has developed a mobile advanced electric reactor (AER) for decontamination of soil. The pyrolytic process uses thermal radiation (near infrared) at 2473-2773 K and has a PCB destruction efficiency of 99.9999%. Operating under reducing conditions minimizes the possible formation of PCDD and PCDF, and lessens the change of explosion. Costs are about the same as that of rotary kiln incineration.

Regarding in situ processes of spill cleanup, Battelle Pacific Northwest (with help from Rockwell International) is testing a soil vitrification process for trapping and destroying PCB. This method uses four large electrodes inserted into the ground surrounding the spill site. A high electrical current heats the ground between and surrounding the electrodes to 1973 K, effectively melting the soil and pyrolyzing, combusting, and vaporizing organics. Released gases are collected and the resultant glassy residue is left in place. In a small-scale laboratory test of 500 mg/kg PCB in soil, the glassy residue had no detectable PCB and there was little PCB immediately adjacent to it (implying that the PCB destruction-volatilization rate was greater than the migration rate). The process generated only 400 ng and 100 ng of PCDF (tetra, penta) and PCDD (hepta, octa), respectively. Destruction efficiencies were 99.95%.

CONCLUSIONS

The previous discussion has indicated that for 100% PCB, Askarel, or Aroclor, there is only one practical method of destruction--incineration at an ANNEX I (or equivalent) EPA approved facility. There are additional options for destruction or decontamination of PCB-contaiminated liquids. EPA has selectively approved chemical dechlorination for PCB concentrations <1%. Under controlled conditions, a high-efficiency boiler is permitted to burn <500 mg/kg PCB concentrations. PCB concentrations <50 mg/kg are not regulated (except for spills near feed sources). Use as road oil is not permitted.

PCB-contaminated soil (resulting from a spill) must be cleaned up. Excavation and removal followed by incineration (ANNEX I) is EPA approved. Other options such as soil washing (extraction) or in situ PCB destruction require demonstration of feasibility before EPA acceptance.

ACKNOWLEDGMENT

The author thanks Gil Addis (EPRI), Ralph Komai (Southern California Gas Company), and Walt Weyzen (EPRI) for helpful discussions and guidance.

REFERENCES

1. J. V. Zbozinek, T. J. Chang, J. R. Marsh, P. K. McCormick, and J. E. Court (SCS Engineers, Inc), "PCB Disposal Manual," Electric Power Research Institute, EPRI CS-4098, Final Report, (June 1985).

2. D. G. Ackerman, L. L. Scinto, P. S. Bakshi, R. G. Dulumyea, R. J. Johnson, G. Richard, and A. M. Takata, "Guidelines for the Disposal of PCBs and PCB Items by Thermal Destruction," United States Environmental Protection Agency, Industrial Environmental Research Laboratory, Research Triangle Park, NC, (February 1981), EPA-600/2-81-022. (Available from NTIS as PB 81-182339).

3. D. S. Duvall and W. A. Rubey, "Laboratory Evaluation of High-Temperature Destruction of Polychlorinated Biphenyls and Related Compounds," United States Environmental Protection Agency, Municipal Environmental Research Laboratory, Cincinnati, OH, (December 1977), EPA-600/2-77-228.

4. R. J. Moolenaar, Distribution and Fate of Chlorobiphenyls in the Environment, in: "Advances in Exposure, Health, and Environmental Effects Studies of PCBs: Symposium Proceedings," R. J. Davenport and B. K. Bernard, eds., Office of Toxic Substances, United States Environmental Protection Agency, Washington, DC, Report

No. LSI-TR-507-137B, NTIS No. PB84-135771 (December 1983), pp. 67-96.

5. M. D. Erickson, "Analytical Chemistry of PCBs," Stoneham, MA:Butterworth Publishers, (1986), Chapter 2, pp. 37-39.

6. P. A. Taylor, "The Impact of How Clean is Clean on Remedial Investigations of PCB-Contaminated Soils (Case Histories and Comparison)," Montech '86 Conference of IEEE, Montreal, PQ., CANADA, (September-October 1986).

COMBUSTION BY PRODUCTS
AND REPLACEMENT LIQUIDS

PRODUCTS OF PYROLYSIS AND COMBUSTION OF LIQUID PCB SUBSTITUTES

Sueann L. Mazer

University of Dayton Research Institute
Environmental Sciences Group
300 College Park
Dayton, OH 45469

NEED FOR PCB SUBSTITUTES

Although PCBs have been used as dielectric fluids since the 1930's, concerns have been raised only relatively recently about the environmental persistance of these compounds, as well as possible adverse health effects attributed to exposure to trace fire byproducts. PCBs were first manufactured commercially in 1929, and were used as transformer and capacitor fluids, heat transfer fluids, plasticizers, and hydraulic lubricants. PCBs have desirable cooling, insulating, and dielectric properties, and hence were considered suitable fluids for use in electrical devices. Because PCBs are relatively nonflammable, they were thought to be ideal for use in transformers and capacitors in indoor locations. Despite these advantages, PCB use has, over the past 15 to 20 years, become restricted due to environmental and toxicological concerns. In 1971, the sole U.S. producer of PCBs began voluntarily reducing the production and distribution of PCBs in response to environmental concerns[1]. In 1977, production of PCBs in the United States ceased entirely.

Since 1979, PCBs have been the target of increasing legislative control. In Canada, they are regulated under the Environmental Contaminants Act. In the United States, they are regulated under the Toxic Substances Control Act and EPA rules promulgated in 1979, 1982, and 1985. The Toxic Substances Control Act (TSCA) basically stated that PCBs could no longer be used except in cases specifically authorized by the EPA. EPA's 1979 rule authorized the continued use of PCBs in enclosed systems, and transformers at that time were considered to be enclosed systems. Later rules on PCB use became more restrictive. The 1982 rule (Electrical Equipment Rule) prohibited the use of PCBs in locations where they might leak into food or feed facilities, and required record-keeping and periodic inspections of PCB-containing equipment. Finally, the 1985 rule (Fire Rule) required that certain PCB transformers found in or near commercial buildings be phased out by 1990, and required other measures be taken to reduce the likelihood of PCB fires and spread of smoke through buildings.

Since 1979, many studies have identified other chlorinated aromatic compounds present in trace levels in the PCBs, or produced as incomplete oxidation products in PCB fires[2-5]. These compounds include polychlorinated dibenzofurans (PCDFs), polychlorinated biphenylenes (PCBPs), polychlorinated pyrenes (PCPYs), and polychlorinated chrysenes (PCCYs). Polychlorinated dibenzodioxins (PCDDs) have also been detected from fires where trichlorobenzenes were present with PCBs as the dielectric fluid[6]. Animal tests and industrial incidents have shown some of the PCDD and PCDF isomers to be much more toxic than the PCBs and trichlorobenzenes themselves[7-13].

Decontamination costs for buildings where PCB fires have occurred have mounted into the millions of dollars. For example, clean-ups in Binghamton, New York; Santa Fe, New Mexico; and San Francisco, California have cost on the order of $28 million, $8.6 million, and $20 million, respectively[14].

In summary, environmental concerns, legislative requirements, potential fire toxicity, and high fire site decontamination costs have all contributed to the need to replace PCBs in electrical equipment.

OVERVIEW OF PCB REPLACEMENTS

Liquid, solid, and gaseous materials have been used as PCB replacements. Some of these materials are suitable for retrofilling existing equipment, while others are manufactured solely for use in new equipment. Solid replacements are used in dry-type transformers, and include epoxy and PVC, while gaseous replacements include chlorofluorocarbons (freons) and sulfur hexafluoride. Liquid PCB replacements, which will constitute the subject of this chapter, are quite diverse, encompassing silicones, high-temperature hydrocarbons, synthetic esters, alkylated aromatics, chlorinated benzenes, and others. A brief summary of liquid replacement fluids is given in Table 1.

Researchers at SCS Engineers have recently published a literature search on combustion and pyrolysis products for many of these fluids[15]. Results of this review are summarized in Table 2, which presents the thermal decomposition products actually observed in laboratory experiments. While the authors of this review do a good job of predicting products where actual data is lacking, it may be observed from Table 2 that laboratory data is indeed lacking in the majority of the cases.

RECENT PYROLYSIS AND COMBUSTION STUDIES OF LIQUID PCB SUBSTITUTES

Introduction

While some pyrolysis and/or combustion products have been identified for about 33% of the PCB replacement liquids, there remain a number of fluids for which there is no data in the open literature. Under a contract from the Electric Power Research Institute (EPRI), researchers at the University of Dayton have been working to fill in some of these information gaps. Thermal decomposition products have now been identified for a number of fluids where previously no information existed. These fluids include capacitor fluids containing n-propylbiphenyl, isopropylbiphenyl, phenylxylylethane, and dixylylethane, as well as a high-temperature hydrocarbon used as a transformer fluid. The registered trade names for the capacitor fluids were Wemcol, Selectrol, Edisol II, Edisol III, and Dielektrol, while the registered trade name for the high-

Table 1. Liquid PCB Substitutes

Silicones (polydimethylsiloxanes)

High-Temperature hydrocarbons

    RTEmp fluid
    Polyalphaolefins [poly (1-octene)]

Esters

    Phthalates
    Pentaerythritol
    Benzylneocaprate

Alkylated aromatics

    Phenylxylylethane
    Dixylylethane
    Isopropylbiphenyl
    n-Propylbiphenyl
    Diisopropylnaphthalene

Chlorinated benzenes

Tetrachloroethylene

Butylated monochlorodiphenylether

Table 2. Summary of Known Thermal Decomposition Products from PCB
Replacements[15]

Replacement Fluid	Thermal Decomposition Products Cited in the Literature
Polydimethylsiloxane	Cyclic siloxanes, Short-chain hydrocarbons, Amorphous silica
RTEmp fluid	No information - see next section
Polyalphaolefins	No information
Phthalates	Phthalic anhydride, Alkenes, Alcohols, Aldehydes
Benzyl neocaprate	No information
Phenylxylylethane	No information - see next section
Dixylylethane	No information - see next section
Isopropylbiphenyl	No information - see next section
n-Propylbiphenyl	No information - see next section
Chlorinated benzenes	Chlorophenols, PCBs, PCDDs, PCDFs, Polychlorinated naphthalenes, Vinyl chloride, Polychlorinated styrenes, Polychlorinated diphenyl ethers, Polychlorinated biphenylenes
Tetrachloroethylene	No information
Butylated monochlorodiphenyl ether	No information

temperature hydrocarbon was RTEmp fluid. A summary of the chemical composition of the fluids as received (some had seen usage) is given in Table 3. These major constituents were identified by gas chromatographic-mass spectrometric (GC-MS) analysis.

Instrumentation and Approach

Thermal decomposition experiments on these fluids were conducted using a highly controlled laboratory test system called the System for Thermal Diagnostic Studies (STDS). With this system, a thermal decomposition unit, which incorporates a high-temperature furnace, is interfaced directly with an analytical system used to separate and detect organic thermal breakdown products. Within this one instrumentation assembly, the sample is therefore vaporized and decomposed at high temperature, and products of this decomposition are identified and quantified. This type of closed system permits tests to be done relatively rapidly and without the errors involved with extractions or desorptions of air sampling traps. Furthermore, conditions are highly controlled and can be reproduced exactly, so results from various fluids can be readily intercompared.

A block diagram of the STDS is shown in Figure 1. A key feature of the STDS is the modular construction which permits customization of any of the components. The sample insertion area, for example, can be used for introduction of gaseous, liquid, solid, or polymeric samples. The high-temperature furnace (up to 1050°C) can be fitted with quartz test cells which are interchangeable and can be customized for the researcher's particular application. A quartz tubular flow reactor was used as the test cell for the experiments described herein. The sample insertion region and reactor are housed within an HP 5890 gas chromatograph (GC). The precise temperature control of the GC permits highly controlled admission of materials into the reactor and quantitative transport of materials throughout the system. The thermal processor is interfaced to a control console which permits precise control of flows, pressures, residence times, temperatures, and gaseous atmospheres within the reactor chamber. Virtually any gas which is noncorrosive to quartz can be used in this system.

Downstream of the thermal processor is an in-line analytical system capable of cryogenic trapping, separation, and detection of thermal decomposition products. For the replacement fluids, the thermal decomposition products were trapped using liquid nitrogen coolant at the head of a capillary GC column housed within an HP 5890 GC. The GC was then used to separate the products, and detection was accomplished using an HP 5970B mass selective detector (MSD). The MSD is a compact quadrupole mass spectrometer which permits analytes to be identified via their fragmentation patterns and quantified via peak areas.

A number of variables are very important in controlling the thermal reaction products which are formed in a fire. Some of these are temperature, time at temperature, and oxygen level. While all three of these can be readily varied with the STDS, permitting data to be generated under many types of conditions, it was desired to obtain only screening data for purposes of this study. Two experiments or tests were performed for each fluid, one in an air (combustion) atmosphere and one in a nitrogen (pyrolytic) atmosphere. Products from the air atmosphere are those likely to form from open burning, while those from the nitrogen atmosphere are those likely to form in sealed equipment or in the localized oxygen-deficient environments which are invariably present in open fires. A reactor temperature of 700°C was chosen for pyrolysis studies,

Table 3. Major Constituents of Capacitor Fluid Samples

Sample Code	Sample ID	Constituents[a]
A	Wemcol	C_3-Biphenyl (2 isomers)
B	Edisol III	C_3-Biphenyl (2 isomers) and Phenylxylylethane
C	Edisol II	C_3-Biphenyl (2 isomers) and Dixylylethane
D	Selectrol	Phenylxylylethane (3 isomers)
E	Dielektrol	Phenylxylylethane (1 isomer)
F	RTEmp fluid	Mixture of many aliphatic hydrocarbons

[a]Other isomeric structures possible.

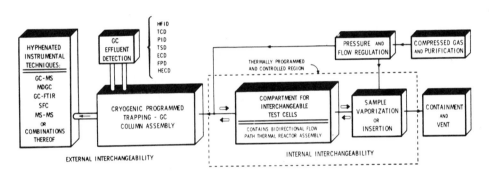

Fig. 1. System for Thermal Diagnostic Studies (STDS).

and 600°C or 625°C for combustion studies. These conditions were chosen as average "worst-case" temperatures in which yields of organic byproducts were maximized.

Experimental Procedures

Figure 2 depicts a block diagram for the thermal decomposition experiments for the PCB replacements. These samples were injected into the STDS as neat liquids, were heated to effect volatilization, and were then swept immediately into the reactor, where they were thermally decomposed in either air or nitrogen at a one second residence time. The reactor effluent, including both undecomposed parent materials and thermal reaction products, was cryogenically trapped at the head of the inline capillary GC column. After the thermal decomposition experiment was complete, products were analyzed using programmed temperature GC-MS. Analysis conditions are given in Table 4.

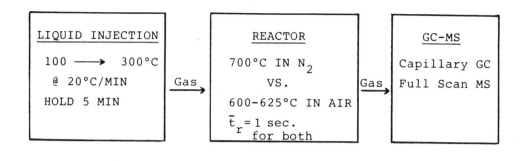

LIQUID INJECTION	REACTOR	GC-MS
100 $\longrightarrow$ 300°C	700°C IN N_2	Capillary GC
@ 20°C/MIN	VS.	Full Scan MS
HOLD 5 MIN	600-625°C IN AIR	

Gas → between LIQUID INJECTION and REACTOR

Gas → between REACTOR and GC-MS

$\bar{t}_r = 1$ sec. for both

| SAMPLE VAPORIZATION | THERMAL DECOMPOSITION | PRODUCT ANALYSIS |

Fig. 2. Block Diagram for the Thermal Decomposition Experiments for the PCB Replacements.

Table 4. GC-MS Conditions

INSTRUMENT: HP 5890 GC & HP 5970B MSD (Quadrupole Mass Spectrometer)

COLUMN: HP-5 Fused Silica Capillary Column, 25 m x 0.2 mm ID, 0.33 μ film thickness

GC TEMPERATURE PROGRAMS: For Capacitor fluids - Hold at -10°C for 1 min; program at 10°C/min. to 300°C; hold for 8 min.

For RTEmp - Hold at -60°C for 1 min; program at 10°C/min. to 300°C; hold for 23 min.

GC-MS TRANSFER LINE TEMPERATURE: 275°C

CARRIER GAS: Helium

IONIZATION MODE: Electron Impact

ELECTRON ENERGY: 70 eV

MASS RANGE SCANNED: 40-500 amu

SCAN SPEED: 1 sec/decade

Results and Discussion

Capacitor Fluids. Figure 3 illustrates some typical data obtained from the thermal decomposition of the five capacitor fluids. This figure shows a chromatogram of the components present in Selectrol, as well as chromatograms of pyrolysis and combustion products. In these

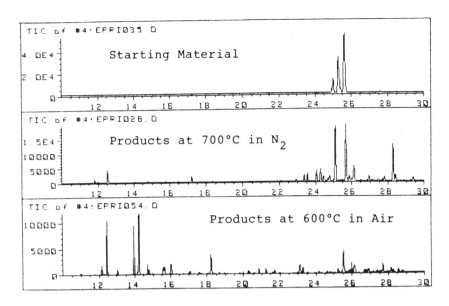

Fig. 3. Total Ion Chromatograms Obtained from Thermal Decomposition Study
Conducted for Selectrol Capacitor Fluid Sample.

chromatograms, each peak represents at least one chemical constituent
detected during the analysis. These typical chromatograms illustrate
that the starting materials were generally mixtures of compounds, and
that the organic pyrolysis and combustion products were much more complex
mixtures. These products were identified by examining the mass spectra
of each peak, and yields were calculated semiquantitatively.

The results of the data analyses described above are given in Table
5 and 6 for pyrolytic conditions and combustion conditions, respectively.
To facilitate comparisons, results from all five capacitor fluids are
provided in each of these tables. While there were some common products
for all five capacitor fluids, the highest-yield products were highly
dependent upon the chemical composition of the starting materials. The
samples composed primarily of propylbiphenyls (Wemcol and the Edisols)
yielded dihydrophenanthrenes (or isomers thereof) as major products in
both air and pyrolytic atmospheres. Biphenylcarboxaldehydes and C_3-
alkenylbiphenyls were other high-yield products in air, while biphenyl
was another high-yield product in nitrogen. The samples containing
primarily phenylxylylethanes (Selectrol and Dielektrol) yielded phenyl-
xylylmethanes, methylanthracenes, and phenylxylylethenes (or isomers of
these three types of compounds) as major products in the nitrogen atmosphere. For these samples, styrene, benzaldehyde, and phenol were common
high-yield products in air. The structures of the highest yield products
from the five capacitor fluids are shown in Table 7.

RTEmp fluid. Unlike the capacitor fluids, which were relatively
simple formulations, GC-MS analysis showed RTEmp fluid to be a very
complex mixture of hundreds of aliphatic hydrocarbons, with most components boiling at about 300°C or higher. Preliminary experiments showed
that in air, this mixture began to thermally decompose at or below 300°C.
Figure 4 shows chromatograms of RTEmp fluid decomposition products obtained at 700°C in nitrogen and 600°C in air. The identities of the

Table 5. Major Thermal Decomposition Products from Capacitor Fluids at 700°C in Nitrogen[a]

GC RETENTION TIME (MIN)	TENTATIVE IDENTIFICATION[c]	% YIELD[b]				
		A[d]	B	C	D	E
3.68	Benzene	7.2	2.1	2.9	1.0	1.1
9.12	Toluene	0.6[e]	0.5	0.5	0.6	0.4
11.83	C_2-Benzene	ND	ND	ND	0.6	0.5
12.53	Styrene	4.1	3.1	2.3	4.0	3.2
14.48	C_3-Alkenylbenzene	0.8	ND	ND	ND	ND
17.12	C_4-Alkenylbenzene or methyldihydroindene	ND	0.8	0.8	1.4	2.0
18.31	Naphthalene or azulene or methylene-1H-indene	2.8	ND	1.5	ND	ND
21.34	Biphenyl	20.4	3.3	8.3	ND	ND
22.81	Methylbiphenyl or diphenylmethane	1.0	ND	ND	ND	ND
23.29	C_2-Biphenyl or phenyl-(methyl phenyl)-methane	ND	1.5	ND	2.0	1.4
23.47	C_2-Biphenyl or phenyl-(methyl phenyl)-methane	ND	0.8	ND	2.6	0.7
23.93	C_2-Biphenyl or phenyl-(methyl phenyl)-methane	1.6	ND	0.7	ND	ND
23.96	C_3-Biphenyl or phenylxylyl-methane	ND	2.5	ND	3.3	4.8
24.20	C_3-Biphenyl or phenylxylyl-methane	ND	2.2	ND	4.3	2.3
24.20	Unknown mixture	5.1	ND	1.6	ND	ND
24.35	C_2-9H-Fluorene	ND	ND	ND	ND	0.5
24.40	Dihydrophenanthrene or diphenylethene	31.1	17.8	21.8	ND	ND
24.45	Unknown mixture	ND	ND	ND	2.1	ND
24.67	C_2-9H-Fluorene	ND	ND	ND	0.8	ND
24.70	Dihydrophenanthrene or diphenylethene	13.4	8.6	10.2	ND	ND
24.77	C_3-Biphenyl or phenylxylyl-methane	ND	ND	ND	1.9	ND
25.00	C_3-Biphenyl or phenylxylyl-methane	ND	ND	ND	19.7	27.8
25.38	C_3-Alkenylbiphenyl	6.0	1.5	4.7	ND	ND
25.82	Phenylxylylethane or C_4-biphenyl	ND	1.2	ND	2.3	1.6
25.83	C_3-Alkenylbiphenyl	2.5	ND	0.8	ND	ND
26.05	Di-(methylphenyl)-ethene or phenylxylylethene or C_4-alkenylbiphenyl	ND	2.9	ND	6.8	10.1
26.49	C_5-Biphenyl or dixylyl-methane	ND	ND	0.6	ND	ND
26.79	Anthracene or phenanthrene	1.7	ND	ND	1.6	ND
27.38	C_2-9H-Fluorene	ND	ND	ND	ND	1.4
27.58	C_5-Biphenyl or dixylyl-methane	ND	ND	3.6	ND	ND
27.69	C_2-9H-Fluorene	ND	ND	ND	1.6	4.6
28.18	Methylanthracene or methylphenanthrene or phenylindene	ND	ND	ND	14.2	1.7

194

Table 5 (Concluded)

28.32	Methylanthracene or methylphenanthrene or phenylindene	ND	1.0	ND	2.3	5.1
28.60	Di-(methylphenyl)-ethene or phenylxylylethene or C_4-alkenylbiphenyl	ND	ND	ND	ND	0.6
29.38	C_2-Anthracene or C_2-phenanthrene	ND	ND	ND	1.3	ND

[a]Listed are chromatographable products boiling at ~80°C or above and formed at yields of ~0.5% or greater. Undecomposed starting materials are not listed as products.

[b]Yields were calculated relative to total peak areas resulting from non-degradative analyses of starting materials. These yields are semiquantitative values accurate to within a factor of five.

[c]Identifications based primarily on evaluation of library search results. Other isomers are possible.

[d]Sample identifications refer to the following: A = Wemcol; B = Edisol III; C = Edisol II; D = Selectrol; E = Dielektrol

[e]ND = not detected above ~0.5% yield.

Table 6. Major Thermal Decomposition Products from Capacitor Fluids at 600°C in Air[a]

GC RETENTION TIME (MIN)	TENTATIVE IDENTIFICATION[c]	% YIELD[b]				
		A[d]	B	C	D	E
3.66	Benzene	ND[e]	ND	ND	1.6	ND
12.43	Styrene	ND	1.7	ND	5.2	3.7
12.13	Unknown containing oxygen	ND	ND	0.5	1.0	ND
13.04	Quinone	ND	ND	ND	0.6	ND
13.97	Benzaldehyde	0.5	2.4	1.2	5.1	3.3
14.24	Phenol	0.6	1.4	1.5	7.8	2.4
14.70	Benzofuran	ND	ND	0.8	1.0	ND
15.65	1H-Indene or C_3-alkynylbenzene	ND	0.6	0.5	0.9	ND
16.02	C_3-Benzene	ND	ND	ND	0.6	0.5
16.02	Methylphenol	ND	ND	ND	0.6	ND
16.02	Methylbenzaldehyde	ND	ND	ND	0.6	ND
17.16	Methyldihydroindene or C_4-alkenylbenzene	ND	ND	ND	ND	0.7
18.23	Naphthalene or azulene or methylene-1H-indene	0.6	0.7	1.0	1.8	0.6
18.23	Naphthalenedione	ND	ND	ND	0.8	ND
18.74	Dimethylbenzaldehyde	ND	ND	ND	ND	0.6
20.82	Indenedione	ND	ND	ND	0.6	ND
21.24	Biphenyl	1.0	0.9	1.8	0.6	ND
23.27	Dibenzofuran	1.1	1.2	1.9	0.8	ND

Table 6 (Concluded)

		A	B	C	D	E
23.12	Naphthalene carboxaldehyde	ND	ND	ND	0.6	ND
24.36	Dihydrophenanthrene or diphenylethene	4.3	4.5	6.9	ND	ND
24.66	Dihydrophenanthrene or diphenylethene	2.5	2.9	4.4	ND	ND
25.35	C_3-Alkenylbiphenyl	2.5	2.5	3.4	ND	ND
25.43	Biphenylcarboxaldehyde	2.5	1.9	4.0	ND	ND
25.71	Biphenylcarboxaldehyde	1.1	1.1	2.0	ND	ND
25.53	Biphenylol	ND	ND	0.7	ND	ND
25.53	Phenylxylylethane or C_4-biphenyl	ND	ND	0.7	ND	ND
25.81	C_3-Alkenylbiphenyl	1.4	1.2	1.9	ND	ND
26.02	C_3-Alkenylbiphenyl	0.5	1.5	ND	ND	ND
26.13	Anthracenone	1.4	0.8	1.3	ND	ND
26.13	Fluorenone	0.6	0.6	0.8	1.6	ND
26.36	Anthracenone	0.5	ND	0.6	ND	ND
26.49	Anthracenone	0.5	ND	ND	ND	ND
26.02	Di-(methylphenyl)-ethene or Phenylxylylethene or C_4-Alkenylbiphenyl	ND	ND	ND	1.2	3.9
26.62	C_3-Biphenyl or phenylxylyl-methane	0.6	ND	0.8	ND	ND
26.89	Methylanthracene or methylphenanthrene or phenylindene	0.7	ND	0.8	ND	ND
27.72	C_4-Biphenyl or phenylxylyl-ethane	ND	0.6	ND	ND	ND
27.70	Unknown Mixture	ND	ND	ND	1.5	1.4
28.13	Methylanthracene or methylphenanthrene or phenylindene	ND	ND	ND	0.7	ND
28.36	C_5-Alkylbiphenyl or dixylylmethane	ND	ND	ND	ND	2.3
28.56	Di-(methylphenyl)-ethene or phenylxylylethene or C_4-alkenylbiphenyl	ND	ND	ND	0.5	2.0

[a] Listed are chromatographable products boiling at ~80°C or above and formed at yields of ~0.5% or greater. Undecomposed starting materials are not listed as products. Samples A, B, and C, were decomposed at 625°C, while samples D and E were decomposed at 600°C.

[b] Yields were calculated relative to total peak areas resulting from nondegradative analyses of starting materials. These yields are semi-quantitative values accurate to within a factor of five.

[c] Identifications based primarily on evaluation of library search results. Other isomers are possible.

[d] Sample identifications refer to the following: A = Wemcol; B = Edisol III; C = Edisol II; D = Selectrol; E = Dielektrol.

[e] ND = not detected above ~0.5% yield.

196

Table 7. Highest-Yield Products from Capacitor Fluids Under Conditions
Tested

Primary Constituents	Trade Names	Combustion Products	Pyrolysis Products

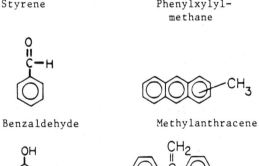

Wemcol
Edisol II
Edisol III

n-Propyl-
biphenyl

Dihydropenanthrene

Dihydrophenanthrene

+

Isopropyl-
biphenyl

Biphenylcarboxalde-
hyde

Biphenyl

C_3-Alkenylbiphenyl

Selectrol
Dielektrol

Phenylxylyl-
ethane(s)

Styrene

Phenylxylyl-
methane

Benzaldehyde

Methylanthracene

Phenol

Phenylxylylethene

197

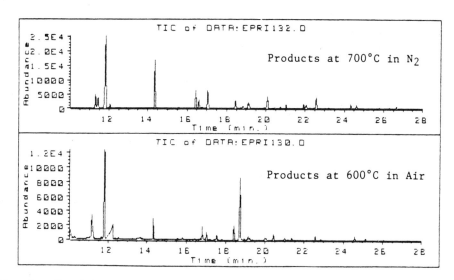

Fig. 4. Total Ion Chromatograms Obtained from Thermal Decomposition Study
Conducted for RTEmp Fluid Sample.

peaks in these chromatograms are given in Table 8 (pyrolysis products)
and Table 9 (combustion products). The highest-yield pyrolysis products
were cyclopentadiene, benzene, and toluene, while the highest-yield
combustion products were benzene, phenol, and acetic acid.

TOXICOLOGICAL EVALUATION OF PYROLYSIS AND COMBUSTION BYPRODUCTS

Chemical analyses of combustion and pyrolysis products from various
PCB replacements would be of little practical value without a discussion
of the toxicology of these products and implications for human health
risks. However, basing a human health risk assessment upon toxicities of
individual thermal decomposition products is in itself a risky venture.
A number of assumptions are made in doing this, two of which are:

1. Results of toxicological tests on animals and microgranisms can
    be extrapolated to humans; and

2. The toxicity of a compound found in a mixture is the same as
    that of the individual pure compound. That is, there are no
    synergistic or antagonistic effects among mixture components.

The reader is cautioned that these assumptions may not hold true in all
cases.

With these caveats in mind, toxicity information will be discussed
for thermal decomposition products from the PCB replacements, with em-
phasis on products which are relatively unique to these utility fluids.
There is a reason for concentrating on unique products. Years of thermal
decomposition research at the University of Dayton have shown that some
stable products are formed from a wide variety of man-made and natural
materials, such as organic industrial wastes, plastics, wood, wool,
fabrics, and sewage sludge. A list of thermal decomposition products
which are very commonly observed is given in Table 10. While these
products are toxic in varying degrees (e.g., benzene is a known human
carcinogen), they are so common as pyrolysis and/or combustion products

Table 8. Major Thermal Decomposition Products from RTEmp fluid at 700°C in Nitrogen[a]

GC RETENTION TIME (MIN)	TENTATIVE PRODUCT IDENTIFICATION[b]	% YIELD[c]
8.41	Cyclopentadiene	0.11
11.30	Cyclohexadiene or methylcyclopentadiene	0.055
11.44	Cyclohexadiene or methylcyclopentadiene	0.038
11.82	Benzene	0.28
12.06	Cyclohexadiene or methylcyclopentadiene	0.014
14.37	Toluene	0.13
16.45	Ethylbenzene	0.050
16.60	Xylene	0.022
17.05	Styrene	0.061
18.46	C_3-Alkylbenzene	0.028
19.09	C_3-Alkenylbenzene or dihydroindene	0.018
19.15	C_3-Alkenylbenzene or dihydroindene	0.014
20.08	Indene or C_3-Alkynylbenzene	0.042
21.03	Methylbenzofuran	0.012
21.92	Methylindene	0.013
22.06	Mixture of 50% naphthalene or isomer & 50% methylindene	0.019
22.57	Naphthalene or isomer	0.038
24.32	Methylnaphthalene	0.014
24.60	Methylnaphthalene	0.012
26.67	Acenaphthylene	0.0099

[a]Listed are chromatographable products boiling at ~60°C or above and formed at yields of ~0.01% or greater. Undecomposed starting materials are not listed as products.

[b]Identifications based primarily on evaluation of library search results. Other isomers are possible.

[c]Yields were calculated relative to starting weight of sample, using anthracene as an external calibration standard for semiquantitatively determining weight of product.

that unless unusually high yields are involved, their presence or absence probably should not constitute the basis for a decision on the safety of a PCB replacement fluid.

Siloxanes, Phthalates and Chlorobenzenes

Table 2 provided data on pyrolysis and combustion products from fluids which had been previously tested. This table indicated that information on partial decompositon products was available in the literature for siloxanes, phthalates and chlorobenzenes.

With the exception of chlorobenzenes, the fluids previously tested had yielded products of low to moderate toxicity. This indicates that fluids such as siloxanes and phthalates may be suitable PCB substitutes, but more extensive testing over a wider range of conditions is recommended to ensure that this is indeed the case.

Table 9. Major Thermal Decomposition Products from RTEmp fluid at 600°C in Air[a]

GC RETENTION TIME (MIN)	TENTATIVE PRODUCT IDENTIFICATION[b]	% YIELD[c]
6.87	Unknown	0.012
9.94	Pentene	0.033
11.14	Unknown	0.048
11.80	Benzene	0.14
12.24	Acetic acid	0.060
14.35	Toluene	0.020
16.82	Unknown, probably unsaturated ketone	0.017
17.04	Styrene	0.0074
17.56	Quinone	0.0058
18.43	Benzaldehyde	0.016
18.74	Phenol	0.089
19.15	Benzofuran	0.0036
20.43	70% Methylphenol & 30% C_3-benzene	0.0080
22.56	Naphthalene	0.0058
24.56	Phthalic anhydride	0.0063

[a] Listed are chromatographable products boiling at ~60°C or above and formed at yields of ~0.004% or greater. Undecomposed starting materials are not listed as products.

[b] Identifications based primarily on evaluation of library search results. Other isomers are possible.

[c] Yields were calculated relative to starting weight of sample, using anthracene as an external calibration standard for semiquantitatively determining weight of product.

Table 10. Some Very Common Thermal Decomposition Products

Benzene	Toluene
C_2-Benzene	Styrene
Naphthalene or isomer	Biphenyl
Anthracene or isomer	Quinone
Benzaldehyde	Phenol
Benzofuran	1H-Indene or isomer
Dibenzofuran	

The chlorobenzenes, on the other hand, had yielded chlorinated products ranging in toxicity from moderate (e.g., chlorophenols) to extremely high (e.g., chlorinated dioxins). Chlorobenzenes are used as dielectrics in combination with other fluids (e.g., PCBs or phthalates). Based upon the teratogenic, tumorigenic and other toxic effects of some of the combustion byproducts of chlorobenzenes, their continued use is not recommended.

Alkylaromatic Capacitor Fluids

The highest-yield products from Wemcol, Edisol II, Edisol III, Selectrol, and Dielektrol were summarized in Table 7. The Registry of Toxic Effects of Chemical Substances (RTECs)[16] was searched for information on the toxicities of these products. Where only generic identifications could be provided from the mass spectral results, the registry was searched for all possible isomers. For example, the "C_3-alkenylbiphenyl" result translated to a search for propenylbiphenyl and ethenylmethylbiphenyl listings. Table 11, which summarizes the status of information available in the registry, shows that toxicological data were available on only about half of the compounds listed in this table. The available information is summarized in Table 12, with comparable information on 2,3,7,8-TCDF and 2,3,7,8-TCDD listed for comparison purposes. The data for the very common thermal decomposition products – benzaldehyde, phenol, and styrene – were not included for reasons mentioned above.

Table 12 indicates that while the toxicities of potential thermal decomposition products from PCBs (TCDF) and chlorobenzenes (TCDD) are on the order of µg/kg, the toxicities of potential high-yield products from the alkylaromatic capacitor fluids are only on the order of mg/kg or

Table 11. Results of Search for Toxicological Data on Highest-Yield Products from Alkylaromatic Capacitor Fluids

FLUID AND ATMOSPHERE[a]	HIGHEST-YIELD PRODUCTS	TOXICITY DATA IN RTECS?[b]
Propylbiphenyls, N_2	Dihydrophenanthrene or	Yes
	Diphenylethene	No
	Biphenyl	Yes
Propylbiphenyls, air	Dihydrophenanthrene or	Yes
	Diphenylethene	No
	Biphenylcarboxaldehyde	No
	C_3-Alkenylbiphenyl	No
Phenylxylylethanes, N_2	Phenylxylylmethane or	No
	C_3-biphenyl	No
	Di-(methylphenyl)-ethene	No
	or Phenylxylylethene or	No
	C_4-Alkenylbiphenyl	No
	Methylanthracene or	Yes
	Methylphenanthrene or	Yes
	Phenylindene	No
Phenylxylylethanes, air	Benzaldehyde	Yes
	Phenol	Yes
	Styrene	Yes

[a]Propylbiphenyls were primary constituents of Wemcol and the Edisol fluids, while phenylxylylethanes were primary constituents of the Selectrol and Dielektrol fluids.

[b]RTECs = Registry of Toxic Effects of Chemical Substances.

Table 12. Toxicity Data Available in RTECS for High-Yield Products
from Alkylaromatic Capacitor Fluids[16]

| COMPOUND | AVAILABLE DATA[a] | | |
	MUTATION	TUMORIGENIC	TOXICITY
1,2-Dihydrophenanthrene		skn-mus TDlo: 72 mg/kg	
Biphenyl	sce-ham: fbr 100 μmol/L	orl-mus TDlo: 56 g/kg scu-mus TDlo: 46 mg/kg	ihl-hmn TClo: 4400 μg/m^3 orl-rat LD50: 3280 mg/kg ivn-mus LD50: 56 mg/kg orl-rbt LD50: 2400 mg/kg
2-Methylanthracene	mma-sat 80 μmol/L/2H		
9-Methylanthracene	mma-sat 75 μmol/L/2H dnd-esc: 10 μmol/L dnd-mam: lym 100 μmol	ipr-mus TDlo: 11 mg/kg	
1-Methylphenanthrene	mma-sat 80 μmol/L/2H msc-hmn: lym 30 μmol/L		
2-Methylphenanthrene	mma-sat 40 μmol/L/2H		
2,3,7,8-Tetrachloro- dibenzofuran			orl-gpg LD50: 5 μg/kg
2,3,7,8-Tetrachloro- dibenzo-p-dioxin		skn-mus TDlo: 62 μg/kg/2Y-1 orl-mus TD: 1 μg/kg/2Y-1	orl-rat LD50: 22.5 μg/kg orl-rbt LDlo: 10 μg/kg orl-gpg LD50: 500 ng/kg

[a]RTECS Abbreviations:

dnd = DNA exchange	mam = mammal
esc = Escherichia coli	mma = microsomal mutagenicity assay
fbr = fibroblast	msc = mutation in mammalian somatic
gpg = guinea pig	cells
ham = hamster	mus = mouse
hmn = human	orl = oral
ihl = inhalation	rbt = rabbit
ipr = introperitoneal	sat = Salmonella typhimurium
ivn = intravenous	sce = sister chromatid exchange
lym = lymphocyte	scu = subcutaneous
	skn = skin

g/kg. This obviously represents a substantial improvement in terms of
human health hazard. The available toxicological information was compiled for the remaining products which were listed in Tables 5 and 6.
The data suggest that the propylbiphenyls, phenylxylylethanes, and
dixylethanes are suitable substitutes for PCBs and chlorobenzenes.
However, the toxicologies of many of the products have not yet been
studied, and while there is no reason to suspect toxicities comparable to
those of certain chlorinated dioxins and furans, further study is
definitely warranted.

RTEmp Fluid

The pyrolysis and combustion products from RTEmp fluid were given in
Tables 8 and 9, respectively. The highest-yield pyrolysis products were
cyclopentadiene, benzene, and toluene, while the highest-yield combustion
products were benzene, phenol, and acetic acid. While these types of
products are moderately toxic, they are by no means unique to this fluid.
Many of these products have been detected in this laboratory as decomposition products from maple wood. In fact, most if not all of the
compounds shown in Tables 8 and 9 have been detected from a variety of
combustion sources, such as coal combustion, waste incineration,
coal/refuse combustion, and automobile and truck engine exhaust[17,18].

Even though RTEmp fluid and the alkylaromatic capacitor fluids were
designed for different uses, it is interesting to compare the yields of
aromatic products from RTEmp fluid (Tables 8 and 9) with those from the
aromatic-based capacitor fluids (Tables 5 and 6). Note that RTEmp fluid,
as an aliphatic starting material, produced lower yields of aromatic
products than did the aromatic starting materials. From the standpoint
of fire toxicity, RTEmp fluid appears to be a good substitute for PCBs.

Conclusions

There is now information available on combustion and pyrolysis
products for eight of the fourteen PCB replacement fluids listed in Table
1. Of the eight materials which have been tested (polydimethylsiloxane,
RTEmp, phthalates, phenylxylylethane, dixylylethane, isopropylbiphenyl,
n-propylbiphenyl, and chlorinated benzenes), the chlorobenzenes appear to
present the greatest human health hazard in the event of fire. This is
primarily due to the formation of chlorinated dioxins, chlorinated
furans, and other chlorinated aromatics during combustion. The other
seven fluids appear to be better choices as PCB substitutes, with the
nonaromatic fluids perhaps posing the lowest health risk.

The fluids which remain to be tested are the polyalphaolefins,
pentaerythritol esters, benzylneocaprate, diisopropylnaphthalene,
tetrachloroethylene, and butylated monochlorodiphenyl ether. Of these,
the chlorinated materials should probably receive the highest testing
priority because of the possibility of forming potentially toxic
chlorinated products. There is no a priori reason to suspect that the
nonchlorinated materials will be converted to products of extreme
toxicity.

Many man-years of effort have been devoted to the replacement of
PCBs in electrical equipment. The urgencies of the need to replace PCBs
- the regulatory pressures, the adverse public opinion, the clean-up
costs, and the health and safety concerns - have at times relegated the
issue of fire byproducts from PCB replacements to a position of secondary
concern. Now, however, data are being acquired in this very important
area, data upon which better decisions can be made. While much work

remains to be done, considerable progress has recently been made in examining the potential fire byproducts of PCB replacements.

ACKNOWLEDGMENTS

This work would not have been possible without the help of colleagues within the Environmental Sciences Group (ESG), as well as support and advice of the Electric Power Research Institute (EPRI), and representatives of various utility companies. Many thanks to John Stalter of the ESG for conducting many of the laboratory thermal decomposition tests; to Wayne Rubey of the ESG for development of the STDS and advice on experimental procedures; to Barry Dellinger for advice on the experimental approach; and to Margaret Bertke for typing and other assistance. Thanks also to EPRI for sponsorship of this work under Contract # RP2028-16; to our Project Officer, Gil Addis, for his advice and support; and to the EPRI advisors for providing guidance, samples, and support in this program.

REFERENCES

1. Interdepartmental Task Force on PCBs, Polychlorinated Biphenyls and the Environment. National Technical Information Service, Publication COM-7210419, 1972.

2. B. Janson and G. Sundstrom, "Formation of Polychlorinated Dibenzofurans (PCDFs) During a Fire Accident in Capacitors Containing Polychlorinated Biphenyls (PCB)," in: Chlorinated Dioxins and Related Compounds, Impact on the Environment, O. Hutzinger, et al., eds., Pergamon Press, Elmsford, NY, 1982, pp. 201-207.

3. C. Rappe, et al., "Polychlorinated Dioxins (PCDDs), Dibenzofurans (PCDFs) and Other Polynuclear Aromatics (PCPNAs) Formed During PCB Fires," Chem. Scripta., 20:56-61, 1982.

4. H. R. Buser, C. Rappe, and A. Gara, "Polychlorinated Dibenzofurans (PCDFs) Found in Yusho Oil and in Used Japanese PCB," Chemosphere, 7:439-449, 1978.

5. J. Nagayama, M. Kuratsune, and Y. Masuda. "Determination of Chlorinated Dibenzofurans in Kanechlors and "Yusho Oil." Bull. Environ. Contam. Toxicol., 15:9-13, 1976.

6. P. E. des Rosiers, B. Westfall, B. Campbell, and A. Lee, "PCB Fires: Preliminary Correlation of Chlorobenzene and PCB Content of the Fluid with PCDF and PCDD Contents of Soot," Proceedings: 1985 EPRI PCB Seminar, March 1986.

7. G. Reggiani, "Toxicology of TCDD and Related Compounds, Observations in Man," in: Chlorinated Dioxins and Related Compounds, Impact on the Environment, O. Hutzinger, et al., eds., Pergamon Press, Elmsford, NY, 1982, pp. 463-493.

8. J. E. Huff, et al., "Long Term Hazards of Polychlorinated Dibenzodioxins and Polychlorinated Dibenzofurans," Environ Health Perspec., 36:221-240, 1980.

9. G. Matthiaschk, "Survey about Toxicological Data of 2,3,7,8-Tetrachlorodibenzo-p-dioxin (TCDD), in: Dioxin, Toxicological and

Chemical Aspects, F. Cattabeni, A. Cavallaro and G. Galli, eds., Spectrum Publications, New York, 1978, pp. 137-141.

10. E. E. McConnell and J. A. Moore. "The Toxicopathology of TCDD," in: Dioxin: Toxicological and Chemical Aspects, F. Cattabeni, A. Cavallaro, and G. Galli, Spectrum Publications, New York, 1978, pp. 137-141.

11. R. A. Neal, P. W. Beatty, and T. A. Gasiewicz. "Studies of the Toxicity of 2,3,7,8-Tetrachlorodibenzo-p-dioxin (TCDD)," Ann. NY. Acad Sci., 320:204-213, 1979.

12. W. P. McNulty, I. H. Pomerantz, and T. J. Farrell, "Chronic Toxicity of 2,3,7,8-Tetrachlorodibenzofuran for Rhesus Macaques," in: Chlorinated Dioxins and Related Compounds, Impact on the Environment, O. Hutzinger, et al., eds., Pergamon Press, Elmsford, NY, 1982, pp. 411-417.

13. J. A. Moore, et al., "Comparative Toxicity of Three Halogenated Dibenzofurans in Guinea Pigs, Mice, and Rhesus Monkeys," Ann. N.Y. Acad. Sci., 320:151-163, 1979.

14. K. C. Ashley, "PCB Decontamination Practices," Proceedings of the Montech '86 IEEE Conference on PCBs and Replacement Fluids, Montreal, Quebec, September 1986.

15. SCS Engineers, "State-of-the-Art Review: Pyrolysis and Combustion of PCB Substitutes," EPRI Publication # EL-4503, March 1986.

16. R. L. Tatken and R. J. Lewis, Sr., eds., Registry of Toxic Effects of Chemical Substances, 1981-82 ed., NIOSH Publication No. 83-107, U.S. Government Printing Office, Washington, D.C.

17. G. A. Junk and C. S. Ford, "A Review of Organic Emissions from Selected Combustion Processes," Chemosphere, 9:187-230, 1980.

18. C. V. Hampton, W. R. Pierson, T. M. Harvey, W. S. Updegove, and R. S. Marano, "Hydrocarbon Gases Emitted from Vehicles on the Road. 1. A Qualitative Gas Chromatography/Mass Spectrometry Survey," Environ. Sci. & Tech., 16:287-298, 1982.

A CRITICAL REVIEW OF SELECTION CRITERIA

FOR PCBs REPLACEMENT LIQUIDS IN POWER APPARATUS

May Carballeira
Laboratoire Central des Industries Électriques
Fontenay-aux-Roses, France

1 INTRODUCTION

PCBs (polychlorinated biphenyls) are widely used in the electrical industry because they have definite advantages over mineral oil, which used to
be the standard insulating fluid for electrical equipment. However,
since PCBs tend to be non-biodegradable and to accumulate in the tissues
of living organisms, most industrialized countries have prohibited the
sale of electrical equipment (transformers and capacitors, both new and
used) containing this chemical. Two problems now remain: what to use
instead of PCBs, and how to dispose of PCBs in a way that does not harm
the environment.

The main objective of this paper is to present a critical review of the
many steps that have led to the selection of replacement fluids for
PCBs. Methods and tests, electrical as well as chemical, physical and
environmental, are suggested and their comparative merits are discussed
for various dielectric liquids.

2 ADVANTAGES OF PCBs

Table 1 compares the properties of mineral oil with those of askarels,
PCB-based transformer fluids. The two main advantages of PCBs are evident: their permittivity is higher (as high as 6 for PCBs used in capacitors) and they have no fire point. The absence of a fire point is due
to:

- the large amount of energy in the bonds of the molecule (ϕ - Cl;
  ϕ - ϕ), which results in a high degree of thermal stability;

- the presence of chlorine in the molecule. When PCBs are subjected to
  high temperatures, they break down and give off hydrochloric acid,
  which considerably reduces the area that might be burned by other combustible products released (Figure 1).

These two advantages have made askarels the ideal dielectric for use in
capacitors, which can be built smaller with no detriment to their efficiency, and in transformers installed in buildings that are inhabited or
frequented by the public.

Table 1

Comparative Properties of Mineral Oil and Askarels[2,3]

CHARACTERISTICS	OIL	TRANSFORMER ASKAREL
RHEOLOGICAL PROPERTIES		
Kinematic viscosity mm^2/s		
40°C	7	8
0°C	–	90
–15°C	90	–
–30°C	270	–
Pour point °C	– 45	– 45
PHYSICAL PROPERTIES		
Density $kg.dm^{-3}$	0.843	1.560
Refraction index at 20°C	1.467	
Refraction index at 25°C		1.614
Boiling point (10^5 Pa)°C	280–400	150
THERMAL PROPERTIES		
Specific heat $J.cm^3.deg^{-1}$		1.67
Thermal conductivity $kJ\ mol^{-1}.h^{-1}$	10 to 14 x 10^{-2}	3.55 x 10^{-3}
CHEMICAL PROPERTIES		
Acid number $mg\ KOH.g^{-1}$	< 0.03	< 0.03
Corrosive sulphur	noncorrosive	
Oxidation stability mg KOH/g		
Acid number	0.08	N/A
Deposit wt%	0.02	
Free chlorine ions ppm	N/A	0.1
Hydrolyzable chlorides ppm	N/A	0.1
ELECTRICAL PROPERTIES		
Permittivity (50 to 60 Hz)		
20°C	2.2	4.6
90°C	2.2	3.5
Dielectric dissipation factor		
90°C	< 2.10^{-3}	< 3.10^{-2}
Breakdown voltage kV (IEC 156)	⩾ 70	⩾ 50
FIRE RESISTANCE		
Flash point °C	158	None
Fire point °C	173	None

In addition, askarels, unlike oils, do not form asphaltene deposits. This is why manufacturers insist that maintenance on PCB-immersed equipment is unnecessary, so much so that international recommendations do not even provide for any aging tests.

With these undeniably attractive features, askarels enjoyed widespread use for 40 years without attracting any attention.

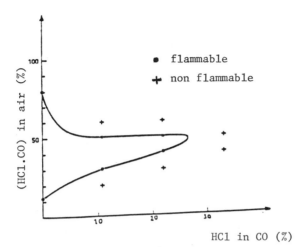

Figure 1. Flammability limits of hydrochloric acid/carbon monoxide mixture.

3 DISADVANTAGES OF PCBs

The disadvantages of PCBs result in large part from the following characteristics:

- Their high permittivity gives them solvent properties that impose caution when selecting the insulation to be immersed in them.

- Their oxidation stability at moderate temperatures and the absence of nitrogen in their structure makes them non-biodegradable.

- The chlorinated vapors given off are toxic and corrosive.

In addition, again because of their well known thermal stability, PCBs are liable to oxidize into polychlorobenzofurans instead of decomposing when exposed to air. Meanwhile, the trichlorobenzenes often found together with PCBs may form polychlorodibenzodioxins, which are well-known to be toxic, carcinogenic and teratogenic. However, this is not the reason for the prohibition of PCBs, as such reactions produce only small quantities of the more toxic substances.[1]

These practical disadvantages, which could perhaps have been eliminated by regulations ensuring proper installation, were amplified by the commercial constraints involved in the use of such materials: direct installation in premises inhabited or frequented by the public, absence of fireproofing and protection, lack of monitoring of the condition of the dielectric liquid, large reductions in insurance premiums, etc.

4 NEW SELECTION CRITERIA FOR DIELECTRIC FLUIDS

4.1 General Remarks

A closer study of Table 1 reveals that the selection criteria for askarels have been more or less modelled on the criteria for oils, with little regard for the particular characteristics of PCBs. In fact, the only differences reside in the allowed amount of free chlorine ions or ionizable chlorine and additives likely to capture the chlorine released.

As for fire-resistance, it is approached solely from the angle of the firepoint. Curiously enough, after several spectacular fires involving plastics, the selection of solid insulation has become much stricter in terms of fire prevention, whereas the standardization of dielectric fluids, at least in Europe, is still in the experimental stage. This is probably because there have been no fatal accidents directly attributable to toxic vapors from burning dielectric fluids.

In fact, standards organizations are working to extend the criteria for solid insulation to dielectric fluids: heat capacity, heat flux given off, limit oxygen index, opacity, vapor corrositivity and toxicity, etc.

At the same time, given the potential market opened up by the banning of PCBs, many dielectric fluids are being proposed, and the selection criteria are undergoing radical changes in terms of both electrical characteristics and fire resistance.

The wide range of available substitutes has enabled the electrical industry to choose a product with the most appropriate electrical characteristics for each application and new selection methods have been adopted.

It should be mentioned that manufacturers had a large amount of data on the compatibility of solid insulation with the fluids used in the past but very little information about new fluids. This must now be compiled.[3]

As for fire resistance, dielectric fluids are studied not only for fire risk but also for the amount of heat given off and the corrosiveness and opacity of the smoke, in the interest of protecting lives and property.

In terms of toxicology and ecotoxicology, the problems experienced with PCBs have made the electrical industry aware that there are physical, management and financial risks and that the problem cannot simply be handed over to the experts.

All these considerations have led the electrical industry to develop new selection methods in addition to the conventional ones, and to find better ways of handling the methods used in less familiar areas such as toxicity in general.

4.2 New Selection Criteria[4]

In addition to the dielectric-strength test at power frequency, which remains the basic test, other criteria have been adopted, such as impulse breakdown voltage, partial-discharge voltage and electrical stability under partial discharges as characterized by gassing.

4.2.1 Impulse breakdown voltage

A knowledge of the behavior of dielectric fluids under impulse conditions has two advantages:

- The effects of the transient voltages to which the fluids will be subjected can be estimated (although the correlation between tests and actual behavior in service is not known);

- It makes it easier to differentiate between the various products, since the impulse withstand voltage is strongly influenced by the chemical composition of the liquid.

In the light of numerous studies conducted by IEC Subcommittee 10A, document 10A(Central Bureau)04, May 1985, two test methods were proposed, as described below.

Method A, derived from ASTM Standard D3300, consists in applying standard positive or negative impulses to the fluid through a system of point-sphere electrodes, increasing the voltage step by step until breakdown occurs. The point has a radius of curvature of 40 to 70 μm; the sphere, a diameter of 12.5 to 13 mm. The distance from point to sphere is 10, 15 or 25 mm, depending on the estimated breakdown voltage of the fluid, which can be determined by a preliminary test if necessary.

Only one impulse is applied to the point at each voltage level, and a pause of at least one minute is allowed between impulses. The fluid must be able to support at least three levels without breakdown. Five impulses must be applied by repeating the test procedure on five different samples. The peak voltage of the impulse that triggers a breakdown is taken as the breakdown voltage. The impulse withstand voltage is defined as the average of the breakdown voltages.

Experience has shown that the value obtained through this method corresponds to a breakdown probability of 10 to 20%.

In Method B, called the "progressive test", impulses of DC voltage are applied and the breakdowns are counted. The results are plotted, and the graph is used to compare the actual probability for the selected voltage with a specified probability, given a known error value.

The sample results cited in Table 2 were obtained with Method A. The main determining parameters were the shape of the impulse voltage, the number of impulses per voltage level, the interelectrode distance and the configuration of the measurement cell (proximity of metallic grounds).

Table 2

Impulse Breakdown Voltage

FLUID	ACC %	NEGATIVE				POSITIVE		
		d mm	V_r kV	E_r kV/mm	t μs	d mm	V_r kV	E_r kV/mm
Naphthenic mineral oil								
A	10	25	142	5.7	21	25	95	3.8
B	18.6	25	134	5.4	30			
C	9.7	25	236	9.4	16	25	93	3.7
D		15	99	6.6		25	85	3.4
Paraffinic mineral oil								
E		25	172	6.9		25	92	3.7
F	5.8	25	160	6.4	24			
G	5.3	25	162	6.5	24			
H		15	100	6.7		25	94	3.8
Silicone oil		15	262	17.5		25	66	2.6
Alkylbenzene		25	242	9.7		25	97	3.9
PCB + trichlorobenzene (TCB)		25	102	4.1				
Trichlorobenzyltoluene(TCBT)		25	138	5.5				
TCBT + TCB		25	108	4.3		25	96	3.8
Phenylxylylethane (PXE)		15	198	13.2		25	160	6.4
Benzyl neocaprate (BNC)		25	92	3.7		25	96	3.8

ACC = aromatic carbon content (IEC 590)
d   = point-to-sphere distance
V_r = breakdown voltage
E_r = V_r/d
t   = time to breakdown

Table 2 shows some breakdown voltages obtained using positive and negative impulses. The breakdown voltages for positive impulses are not very
selective and do not appear to depend on the nature of the fluid, whereas
those obtained from negative impulses seem to closely reflect the components of the fluid. Consequently, dielectric fluids can be roughly
divided into two groups:

- $E_r < 10$ kV/mm: benzylneocaprate, trichlorobenzene, PCB, mineral oil
  and trichlorobenzyltoluene

- $E_r > 10$ kV/mm: alkylbenzene, phenylxylene ethane, silicone oil.

It can be seen that the presence of chlorine atoms (and carbonyl groups)
generally tends to lower the negative impulse voltage -- a phenomenon
confirmed by the fact that the addition of trichlorobenzene tends to
further diminish the performance of trichlorobenzyltoluene, and probably
that of PCB.

4.2.2 Partial-discharge threshold

There is no standard governing this method as yet. The threshold voltage
(U_s) at which partial discharges (PD) appear is the AC voltage at which
PDs begin to achieve the steady state in a given system.

Figure 2 shows a test cell for measuring this threshold voltage. It contains a needle with a point radius of 3 μm. The interelectrode distance
is 70 mm. The cell is airtight, with several bushings used to control
the atmosphere.

The cell is subjected to AC voltage (preferably 50 or 60 Hz) raised at a
uniformly increasing speed (1 kV.s^{-1}). When the first discharge is seen
on the oscilloscope screen, the voltage increase is stopped and repeti-
tion rate of the discharges is recorded. The threshold 20 PD/min and an
amplitude higher than 500 pC are maintained. The amplitude value was
deliberately set high to eliminate any PDs due to the passing of particles between electrodes. Table 3 shows some U_s values for various
fluids under the given test conditions.

The classification according to threshold voltages is not the same as for
impulse voltage test. Again, two groups can be identified:

- $U_s < 60$ kV: silicone oil, polybutane, mineral oil, benzyl neocaprate,
  alkylbenzene

- $U_s > 60$ kV: trichlorobenzene, trichlorobiphenyl, alkylnaphthalene,
  ethane phenylxylene.

The products in the first group have a large aliphatic component, whereas
those in the second group are much more aromatic. The presence of substitute chlorine atoms does not seem to have any harmful effects in this
case.

Table 3

Partial–Discharge Thresholds

Fluid	U_S (radius = 3 μm) kV
Silicone oil (dimethylpolysiloxane)	38.1
Polybutene 1	39.7
Alkylbenzene	41.5
Paraffinic mineral oil (with flow improver)	46.1
Polybutene 2	48.4
Benzyl neocaprate	52.5
Naphthenic mineral oil 1	54.4
Naphthenic mineral oil 2	54.4
Trichlorobenzene	> 80
Trichlorobiphenyl	> 80
Alkylnaphthalene	> 80
Ethane phenylxylene	> 80

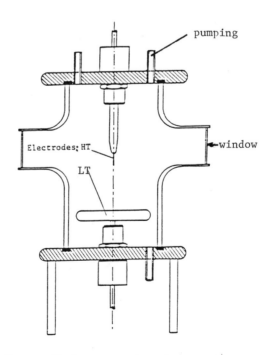

Figure 2. Test cell for measuring the partial–discharge threshold voltage.

4.2.3 Gassing

The term "gassing" refers to the tendency of a liquid subjected to partial discharges in a gas phase to absorb or release gaseous products resulting in a variation in the volume of the phase. IEC document 628 (1985) describes two methods for determining gassing.[5]

Method A can be used to calculate G, the rate at which gases are formed or absorbed in l/min, over a short period (2 h). The test, which is conducted in hydrogen at 80°C and 10 kV, is more suitable for synthetic fluids used in closed systems such as cables, sealed transformers and capacitors.

Mehtod B, which determines the quantity of gas formed or absorbed over a longer period, also serves the purpose of an endurance test. It is more suitable for open equipment such as power transformers. It calculates the volume G of nitrogen gas absorbed or formed after 18 h at 80°C and 12 kV.

Tables 4 and 5 give examples of values obtained with Method A from naphthenic (N) or paraffinic (P) mineral oil, with different aromatic carbon contents and on different dielectric fluids for capacitors.

Again, these tables show that hydrocarbons with pronounced aromatic structures yield larger quantities of gas, although a direct correlation is not possible (in the case of oils).

4.3 Fire Resistance

Here too, the criteria for evaluating the fire resistance of liquid dielectrics are changing drastically because of a new awareness of the need to assess fire hazards as objectively as possible.

In addition to the usual criteria of flash point, fire point and limit oxygen index, some new criteria focusing more on the protection of lives and property have been developed.[6]

For this reason, attention is now being accorded to the notions of combustion time, heat released, and the opacity, corrosivity and toxicity of smoke. There are many ways to assess these risks but one of the most comprehensive methods is the one developed in the U.S. by A. Tewarson of Factory Mutual Research Corporation,[7] which was the object of a proposal in IEC publication 10B, France 38, May 1984. In its present form, the method does not take account of the corrosivity or toxicity of smoke, but its scope could always be broadened to correct this.

4.3.1 Heat released and opacity of smoke

Table 6 contains the results obtained by the Tewarson method[8] on some of the dielectric liquids currently on the market. However, care must be exercised in the use of these data, since they are completely dependent on the operating conditions chosen and any variation could result in inverted classifications. Other calculation methods have also been proposed.[9]

Table 4

Gassing of Mineral Oil

Type of oil	Aromatic carbon (%)	G (Method A)	G (Method B)
N	5.8	+ 22.1	+ 7
	13.4	+ 9.3	+ 7
	16	− 24.6	− 3
	17.2	− 7.9	− 0.3
P	7.9	− 1.9	+ 5
	9.6	− 35.7	− 2.9
	13.8	− 22.9	− 0.5
	17.8	− 34	− 2.7

Table 5

Gassing of Various Fluids (Method A)

Fluid	G (μl/min)
DOP/DEHP	− 63
TRICHOLOROBIPHENYL	− 105
BNC	− 166
DOP + TCB	− 187
TCB	− 190
PXE	− 215
DIPN	− 235
MIPB	− 291

Table 6

Tewarson tests

Reference	Inflammation time*	End of* combustion	Weight (g)	Residue (g)	Weight loss max. (g/min)	Max.O_2 consumption (%)	Max.rate of heat release (kW/m²)	Rate of heat released by convection (kW/m²)	Total heat released (kJ)	Heat by weight (kJ/g)	Max.optic density of smoke (D.O.)
Mineral oil	1'20"	3'25"	8.9	0	10.8	1.8	2150	620	270	30.4	2.6
Piralene	2'	14'	14.75	0	3.6	0.09	105	62	35	2.37	0.35
Ugilec T	2'35"	ends at 4'30"	14	6,9	3.4	0.13	150	90	18	1.3	0.43
Formel NF	-	-	16	0	9.2	0	0	0	0	0	0
Silicone	5'40"	30'	9.6	5,6	0.25	0.07	82	52	140	14.6	0.08
Midel (ester)	3'50"	6'15"	9.3	0	10.9	1.5	1730	640	240	25.8	0.50
Envirotemp 100	4'02"	6'10"	9.5	0	12.3	1.9	2200	960	285	29.9	0.35
Baylectrol 4900	1'15"	2'50"	10.2	0	14.9	1.4	1630	230	215	21.1	3.66
PXE NPCC	1'14"	2'50"	9.6	0	13.4	1.24	1450	250	225	23.5	4.15

*: ' : min
   " : sec

217

A study of these results once again points up the real advantages of chlorinated liquid dielectrics in terms of fire resistance: indeed, they give off the least heat per unit weight and their smoke is the least opaque.

Silicone oil ranks in the middle as to the amount of heat released and the opacity of its smoke. With other methods, however, opacity is very great (silica dust) and the tests conducted to scale on real equipment did not yield the expected results.

The other products are very similar to mineral oil and can produce very opaque smoke (Baylectrol 4900 and PXE NPCC). However, it should be noted that Envirotemp 100 gives off smoke of low opacity.

4.3.2 Corrosivity of smoke

Again, methods are currently being standardized and consequently very little data has been published so far. A draft standard[10] exists in France which is not a part of any IEC publication and which is being extended to cover dielectric liquids. This purely conventional method consists in measuring the pH of water through which gases from pyrolysis of a test sample at 800°C in air are bubbled. The lower the pH, the higher the potential corrosivity. Table 7 gives some of the results for four commercial liquids.

Table 7

Potential Corrosivity of Commercial Liquids

Liquid	Loss of weight %	H_3O+ concentration Mol/l.g	pH for 1 g of pyrolyzed matter
Midel 7131	100	100×10^{-6}	4.0
Shell Diala D mineral oil	100	30×10^{-6}	4.53
604 VS 50 silicone oil	62	18×10^{-6}	4.74
Ugilec T	100	28×10^{-6}	1.55

It is clear that the pyrolysis of Ugilec T, or any chlorated product, releases acid smoke liable to cause corrosion of material; however, it should be remembered that this smoke also limits the spread of fire.

Silicone oil deposits 38% noncorrosive but troublesome silica dust on electrical contacts, along with the corrosion products.

5 TOXICITY AND ECOTOXICITY

Two aspects of these factors must be considered: first, the properties of the product itself, which are always carefully studied before new molecules can be put on the market; and second, the properties of the decomposition products, which have only recently become a recognized factor.[11] Nevertheless, it is still important to know all the physicochemical properties (solubility, vapor pressure, etc.) of the substance being tested.

In the first case, electrical engineers, who are the users of synthetic products, cannot help but take into account the results of studies and respect the regulations in force. Although it may be easy to reproduce an official certificate for a physicochemical or electrical measurement, such is not the case for measurements related to toxicity or ecotoxicity.

In the second case, data and even experimental methods are practically inexistent. The electrical industry is therefore at a loss to solve this problem, which reaches beyond its usual range of experience.

Actually, methods for studying toxicity and ecotoxicity are described in the Official Journal of the European Communities, ref. L 251, 27th year, September 19, 1984 (84/449 EEC) on the classification, packaging and labelling of dangerous substances. It is not our intent here to report the entire contents of this directive (although it would be highly edifying), but merely to give the reader an idea of the properties to be considered and the methods used to determine them (Table 8).

5.1 Methods for Determining Toxicity

The toxic properties taken into account are shown in Table 8. A distinction is made between acute toxicity, subacute toxicity and mutagenicity.

5.1.1 Acute toxicity

This comprises adverse effects occurring within a given time (usually 14 days) after administration of a single dose of a substance.

The first three methods (B1, B2 and B3) are used to determine the median lethal dose (LD_{50}) administered orally or dermally, and the median lethal concentration (LC_{50}).

LD_{50} is the single dose that can statistically be expected to cause the death of 50% of the animals during a specified period. It is expressed as the weight of test substance per unit weight of test animal ($mg.kg^{-1}$).

LD_{50} is the concentration of a substance that can statistically be expected to cause death during exposure or within a fixed time after exposure in 50% of animals exposed for a specified time. LC_{50} is expressed as the weight of test substance per standard volume of air ($mg.l^{-1}$). The exposure time is usually 4 h.

Methods B4, B5 and B6 are designed to determine skin and eye irritation and skin sensitization.

Skin irritation is the occurrence of reversible inflammations after application of a substance.

Table 8

PART A: METHODS FOR THE DETERMINATION OF PHYSICOCHEMICAL PROPERTIES

A. 1.   Melting point/Melting range
A. 2.   Boiling point/Boiling range
A. 3.   Relative density
A. 4.   Vapor pressure
A. 5.   Surface tension
A. 6.   Water solubility
A. 7.   Fat solubility
A. 8.   Partition coefficient
A. 9.   Flash point
A. 10. Flammability (solids)
A. 11. Flammability (gases)
A. 12. Flammability (substances and preparation which, in contact, with water or damp air, evolve highly flammable gases in dangerous quantities)
A. 13. Flammability (solids and liquids)
A. 14. Explosive properties
A. 15. Auto-flammability (determination of the self-ignition temperature of volatile liquids and of gases)
A. 16. Auto-flammability (determination of the relative self-ignition temperature
A. 17. Oxidizing properties

PART B: METHODS FOR THE DETERMINATION OF TOXICITY

GENERAL INTRODUCTION

B. 1.   Acute toxicity (oral)
B. 2.   Acute toxicity (inhalation)
B. 3.   Acute toxicity (dermal)
B. 4.   Acute toxicity (skin irritation)
B. 5.   Acute toxicity (eye irritation)
B. 6.   Acute toxicity (skin sensitization)
B. 7.   Subacute toxicity (oral)
B. 8.   Subacute toxicity (inhalation)
B. 9.   Subacute toxicity (dermal)
B. 10. Other effects − Mutagenesis (_in vitro_ mammalian cytogenetic test)
B. 11. Other effects − Mutagenesis (_in vitro_ mammal bone marrow cytological test, chromosomal analysis)
B. 12. Other effects − Mutagenesis (micronucleus test)
B. 13. Other effects − Mutagenesis (_Escherichia coli_ − reverse mutation assay)
B. 14. Other effects − Mutagenesis (_Salmonella typhimurium_ − reverse mutation assay)

PART C: METHODS FOR THE DETERMINATION OF ECOTOXICITY

C. 1.   Acute toxicity for fish
C. 2.   Acute toxicity for _Daphnia_
C. 3.   Degradation − biotic degradation: modified OECD screening test
C. 4.   Degradation − biotic degradation: modified AFNOR test NF T 90/302
C. 5.   Degradation − biotic degradation: modified Sturm test
C. 6.   Degradation − biotic degradation: closed-bottle test
C. 7.   Degradation − biotic degradation: modified MITI test
C. 8.   Degradation − biochemical oxygen demand
C. 9.   Degradation − chemical oxygen demand
C. 10. Degradation − abiotic degradation: hydrolysis as a function of pH

Eye irritation is the occurrence of reversible effects after application of the substance inside the eye.

In both cases, doses and exposure times are fixed and observations and measurements are made after 1, 24 and 72 h. Skin sensitization (allergic contact dermatitis) is an immunologically medicated cutaneous reaction.

In the case of skin sensitization, the following method is used. The animals are first subjected to the substance under test (induction period). Two weeks after the last induction period, they are given a challenge exposure to determine whether a state of hypersensitivity has been induced. Sensitivity is determined by a study of the skin response to the challenge exposure. The induction injections are intradermic, in the shoulder area, while the challenge exposure is administered by applying a dressing containing the test substance.

Observations are made at specified times.

5.1.2 Subacute or subchronic toxicity

Subacute toxicity comprises the adverse effects occurring in test animals given daily doses of a substance or exposed daily for a brief period, in relation to their life expectancy.

The test substance is administered daily for 28 days either orally, cutaneously or through inhalation. During this period, the animals are examined every day for toxic effects. Blood tests and biochemical measurements are taken at the end. All the animals, including any that have died during the testing, are necropsied.

During this test, the highest dose (or concentration) should be generally non-fatal, while the lowest should not result in any toxic effects. Under ideal conditions, the average dose should produce the minimum observable toxic effects.

5.1.3 Mutagenicity

A preliminary assessment of the mutagenic potential of a substance requires information on two mechanisms:

- chromosomal abnormalities, studies by means of in vivo or in vitro cytogenetic tests on the cells of mammals (B10, B11, B12);

- gene mutations, studied on procaryotes such as Salmonella typhimurium or Escherichia coli.

(a) Chromosomal abnormalities

Short-term tests are conducted to detect chromatinic or chromosomic aberrations and to determine the number of abnormal metaphases.

B10 is a test performed in vitro on cultures of established cell lines. After exposure to the test substance, the cultures are treated with colchicine to accumulate the cells in a metaphase-like stage of mitosis (the chromosomes are then well differentiated). The preparations are then stained and the metaphase cells analysed to determine the mitotic index and the type and number of chromosomal abnormalities.

Test B11 is performed _in vivo_ on mammals exposed to the test substance. Animals are killed at regular intervals, after treatment with colchicine. Samples of their bone marrow are taken and the chromosomes are studied as described above.

B12, known as the "micronucleus test", is also for the study of chromosomal damage. It is based on an increase in the number of micronuclei in the polychromatic erythrocytes of animals treated with the test substance in relation to a control group.

Micronuclei are made up of chromosomes or chromosome fragments lagging in mitosis. When the erythroblasts in the hematopoietic organs are transformed into erythrocytes (red blood corpuscles), the main nucleus is expelled while the micronucleus may be conserved in the cytoplasm. In this test, young polychromatic erythrocytes from the bone marrow of laboratory animals are used. The dose administered is the maximum tolerated, or enough to cause a modification in the ratio of polychromatic to normochromatic erythrocytes.

(b) Gene mutations

Two methods have been proposed: B13 and B14, performed on either Escherichia coli or Salmonella typhimurium.

Both methods consist in measuring the bacteria's reverse mutation in relation to its capacity to synthesize a particular amino acid needed for its development. The + and − signs beside the abbreviation for each amino acid represent the microbe's ability (+) or inability (−) to synthesize that amino acid and thus to develop in a minimal environment deprived of amino acid.

With Escherichia coli, the amino acid considered is thyptophan: trp^l −− trp^+ under the influence of the substance being tested for mutagenicity.

With Salmonella typhirium, the amino acid considered is histidin: His^- −− His^+. These microbial transformations translate the modification of the organism's genome under the influence of mutagenic substances. The test results are expressed as the number of reverse-mutating colonies per treated culture, in relation to control cultures.

5.2 Carcinogenic and Teratogenic Effects

The EEC document does not mention any particular method for assessing carcinogenicity and teratogenicity (tendency to cause developmental malformations or monstrosities).

These risks are assessed from the data obtained through testing for mutagenicity, or through statistical observations on the occurrence of tumors or malformations in humans. The Official Journal of the European Communities (ref. L257, September 16, 1983) classifies carcinogenic and teratogenic substances according to the available data.

5.3 Ecotoxicity

The methods for determining ecotoxicity (Table 8) take account of two aspects: first, direct toxicity for a given species in an ecosystem (C1 and C2); and second, biotic degradation (or biodegradability) and abiotic degradation, the lack of which can cause the substance to become more concentrated as it progresses along the food chain, possibly until it reaches dangerous concentrations (C3 and C10).

5.3.1 Toxicity

(a) Acute toxicity for fish

This test is conducted on zebra fish (Brachydanio rerio). Acute toxicity is expressed as the median lethal concentration (LC_{50}), that is, the concentration (mg.1) required to kill 50% of the fish exposed for a specific duration (preferably 90 h). After exposure, the deaths are recorded every 24 h and the LC_{50} is calculated statistically after each observation.

(b) Acute toxicity for Daphnia

This test measures the median effective concentration (EC_{50}), that is, the concentration ($mg.1^{-1}$) required to produce immobilization. This corresponds to the initial concentration required to immobilize 50% of the daphnids exposed to the substance for 24 h.

5.3.2 Biotic degradation

Methods C5 to C8 are similar in principle if not in practice, being designed to measure the biodegradability of nonvolatile, water-soluble organic compounds in aqueous and aerobic environments at varying initial concentrations. C9, however, is a purely chemical method. The means of assessing biodegradability differs according to the test:

- C3, C4: disappearance of dissolved organic carbon (DOC);

- C5: measurement of actual CO_2 production in relation to the theoretical value ($ThCO_2$) based on the carbon content of the test substance;

- C5, C7, C8: biochemical oxygen demand (BOD). Degradation is defined as the ratio between the BOD and the theoretical oxygen demand (ThOD);

- C9: chemical oxygen demand (COD), a measurement of the oxidation properties of a substance and defined as the quantity of oxygen in an oxidizing reagent consumed by the substance under predetermined conditions.

5.3.3 Abiotic degradation: hydrolysis as a function of pH (C10)

Hydrolysis is one of the principal reactions controlling abiotic degradation (disappearance of non-biodegradable substances into the environment). Since it is a pseudo first-order reaction (large surplus of water), the half-life duration is independent of the concentration, which facilitates extrapolation to environmental conditions.

The method consists in determining the hydrolytic behavior of chemicals with pH values commonly found in the environment (pH 4 to 9).

The substance is dissolved in a large amount of water with controlled pH and temperature. The decrease in concentration over time is monitored. When log [C] is plotted as a function of time, the result should be a straight-line curve (first order) whose slope k_{obs} is the reaction rate constant. Extrapolations can be made for different temperatures using the Arrhenius equation. The unit for the constant k_{obs} is $(time)^{-1}$.

5.4 Toxicity of Decomposition Products

Very little literature is available on the toxicity of the decomposition products of insulating liquids; there are no standard methods, and data are rare and often contradictory. Two approaches are possible:

- Analytical study, which consists in analysing the decomposition products at various temperatures in order to determine a toxicity index. Such a method is already being employed in France for solid insulation (AFNOR test NF C 20 454), and could probably be extended to include liquids. However, it has the disadvantage of providing information on only a few of the possible toxic products (HCl, CO, CO_2, HCN, etc.). Moreover, it disregards the intermediate products of oxidation, some of which are thought to be highly toxic, such as acrolein, polychloro-dibenzofurans and polychlorodibenzodioxins, and it ignores the synergetic effects that may result from exposure to a complex mixture of these products;

- Comprehensive study, which consists in directly exposing rats to vapors produced by pyrolysis. Such a method is described in the VIth EEC report on health and safety in mines and extractive industries. It consists in dropping the liquid on a metal plate heated to 700°C and exposing rats to the vapors produced. Table 9 shows some of the data obtained through this method as reported by CERCHAR.

Table 9

Toxicity of Decomposition Products
Data from CERCHAR Report

Test according to VIth EEC report
(Safety in mines and extractive industries)

Dielectric liquid	Rating	Number of animal mortalities
Mineral oil	10	6/6
Dimethylpolysiloxane	10	6/6
Aliphatic ester of pentaerythritol	10	6/6
Ugilec T	2	0/6

5.5 Translation of Toxicological Data for Classification, Packaging and Labelling of Dangerous Substances

The regulations governing the classification, packaging and labelling of dangerous substances are stated in the Official Journal ref. L 257, 26th year, September 16, 1983, p. 13 and following.

Again, rather than describe the content of all the regulations here, we shall simply explain the principle used in classification.

Substances and preparations are generally classified according to the product's toxicity, expressed as LD_{50} or LC_{50}. Symbols are used to indicate danger, accompanied by R phrases describing the danger and S phrases advising precautions.

Table 10 shows the levels of toxicity according to LD_{50} (for ingestion and skin contact) and LC_{50} (for inhalation), as well as the corresponding symbols and R phrases. Some of the published LD_{50} values are shown in Table 11.

Table 10

European Economic Community
Classification and Labelling of Dangerous Substances
Toxicity Criteria

Category	Symbol	LD_{50} ingestion rat $mg.kg^{-1}$	LD_{50} skin contact rat or rabbit $mg.kg^{-1}$	LC_{50} inhalation rat $mg.l^{-1}.4h^{-1}$
Highly toxic	Skull and crossbones	< 25 (R28)	⩽ 50 (R27)	⩽ 0.5 (R26)
Toxic	Skull and crossbones	25 to 200 (R25)	50 to 400 (R24)	0.5 to 2 (R23)
Harmful	St. Andrew's cross X_n (n = noxious)	200 to 2000 (R22)	400 to 2000 (R21)	2 to 20 (R20)

Dangers of acute or subacute toxicity, mutagenesis, carcinogenesis and teratogenesis are also pointed out through the appropriate R phrases and symbols, for example:

Category	Symbol	R phrase
Harmful substances	X_n	R 40 – Possible risk of irreversible effects
Harmful substances		R 48 – Danger of serious damage to health by prolonged exposure
Irritants	X_i	R 38 – Irritation to skin
		R 36 – Irritation to eyes
		R 41 – Risk of serious damage to eyes

225

Table 11

Published Toxicity Values for Dielectrics

Dielectric liquid	LD_{50} ingestion (rat) $mg.kg^{-1}$	LD_{50} skin contact (rat or rabbit) $mg.kg^{-1}$	Observations
PCB	2 000 to 19 000	800 to 3200	Tolerable limit
TCB	-	-	
T Temp	>10 000		
Midel	>10 000		
Mineral oil	>10 000		
Ugilec T	2 556	2000	Non-classified
Jarylec	> 3 000		
Silicone	>10 000		

6 CONCLUSION

Many replacement liquids for PCBs are now available for both capacitors and transformers.

It should first be remembered that, whatever the qualities of the replacement liquid chosen, the user must conduct tests to ensure compatibility with the other materials used in manufacture (metals, insulation, enamel, varnish, glue), as well as performance tests on models and real equipment. These tests do not fall within the scope of this paper. In addition, changes in operating conditions may result in wrong classification.

For capacitors, the liquids must have a low dielectric dissipation factor, high permittivity or a capacity to tolerate a large average voltage gradient, a good negative impulse voltage, a high partial-discharge threshold, and good gas absorption properties. Esters (MIDEL, MIXOFLEX, etc.), silicone oils (Rhône Poulenc, Dow Corning, etc.) and aromatic hydrocarbons (Jarylec C 101, PXE, DIPN, MIPB, etc.) all meet these criteria according to the type of application.

As for transformers, the choice is more difficult. Halogenic hydrocarbons (Ugilec T, FORMEL, etc.) still offer the greatest safety in terms of fire resistance. However, they have the disadvantage of releasing corrosive vapors under pyrolysis.

Silicone oils are an intermediate stage in that they burn rather badly but without giving off corrosive vapors; also, they have a relatively low thermal capacity but release silica dust.

The other products (esters, heavy aliphatic hydrocarbons, organic esters) have higher fire points than mineral oil but similar heat-producing capacity, and tend to give off dense smoke.

In short, the choice of a dielectric liquid is neither simple nor risk-free. It is particularly important not to neglect the administrative risk. A product is chosen on the basis of current knowledge and legislation but in 30 years (the service life of a piece of electrical equipment) many factors can change.

Finally, the choice of a liquid dielectric is always the fruit of compromise between technical and economic imperatives and the safety of lives and property. It must therefore be remembered that appropriate regulations for installation and operation, oil pits, fire walls, fire-proofing, smoke removal system, absence of overloading, electrical protection, detection of latent defects) can go a long way towards remedying the disadvantages inherent in an imperfect solution.

REFERENCES

1. A. FOURNIE: L'accident de Reims, RGE 1/86, January 1986.

2. IEC Publication 296: Specification for unused mineral insulating oils for transformers and switchgear.

3. IEC Publication 588: Askarels for transformers and capacitors.
    IEC Publication 588-5 and 588-6: Askarels for transformers and capacitors.

4. B. FALLOU, J. PERRET, J. SAMAT, P. VUARCHEX: CIGRE 1986 (15-10) Évolution des critères de sélection des liquides isolants.

5. IEC Publication 628: Gassing of cable and capacitor insulating oils under electrical stress and ionization.

6. UTE draft standard C 27-100: Classification des diélectriques liquides d'après leur comportement au feu.

7. Factory Mutual Research, approval standard: "Less flammable transformer fluids," May 30, 1979.

8. Rhône Poulenc internal study.

9. Ph. MALLET, LCIE Internal   Report, October 1985.

10. UTE draft standard C 20-453.

11. Official Journal of the European Communities L 251, 27th year, September 19, 1984.

CONTRIBUTORS

May Carballeira
LCIE
Fontenay-aux-Roses, France

Gaétan Carrier
Hydro-Québec
Montréal, Qc, Canada

Robert Grob
École Nationale Supérieure de Chimie
Toulouse, France

Jacques Guertin
EPRI
Palo Alto, CA, USA

Jacques Mathieu
École Nationale Supérieure de Chimie
Toulouse, France

Donald Mackay
University of Toronto
Toronto, Ont., Canada

Sue Ann L. Mazer
University of Dayton Research Institute
Dayton, OH, USA

Douglas E. Metcalfe
Canviro Consultants
Waterloo, Ont., Canada

Ross J. Norstrom
Environment Canada
Ottawa, Ont., Canada

Sally Patterson
University of Toronto
Toronto, Ont., Canada

H. Ricau
École Nationale Supérieure de Chimie
Toulouse, France

Stephen Safe
Texas A & M University
College Station, TX, USA

Richard L. Wade
Med-Tox Associates Inc.
Tustin, CA, USA

Ian Webber
Envirocon Systems Inc.
Getzville, NY, USA

John P. Woodyard
International Technology Corporation
Torrance, CA, USA

Enzo M. Zoratto
International Technology Corporation
Pittsburgh, PA, USA

George Zukovs
Canviro Consultants
Waterloo, Ont., Canada

INDEX

Askarels (see Properties)

Biologic
  accumulation, 87-94
    effects on human, 52-58, 76-79,
      108
    effects on wildlife, 85-98,
      108-113

Chromatography
  gas, 35-48
    colums, 36
    detectors, 38
    injectors, 38
    solvants, 40
  results, 38-42, 86-98, 191-198

Decontamination
  after fires, 76-80, 165-168
  guidelines, 101-113
  of concrete, 117-134
  spills, 181
  techniques, 128-132

Destruction, 175-183
  thermal, 177-180
  nonthermal, 180-181

Disposal
  of fluids, 168-170
  of transformers, 142

Ecotoxicity, 223-225

Environmental
  contamination, 101-115
  exposures, 60
  levels, 105-112

Exposure
  assessments, 104, 22
  dermal, 106, 222-225
  environmental, 60
  ingestion, 106, 222-225
  occupational, 59

Fire (see also Smoke and Soot)
  PCDF in, 72-76
  resistance, 215-220
  risks, 76-80
  toxicity, 195-198
  transformers, 140-143

Health (see also Biologic and
Toxicity)
  decontamination, 76-78, 101-113
  human, 55-58, 76-79

Mass spectrometry, 43

Properties
  chemical, 3-30, 208
  electrical, 3-30, 208-215
  physical, 3-30, 51-53, 208
  of PCB
    congeners, 5, 85
    extraction, 46-48
    isomers, 5, 85
    leaching, 155-156
  of PCDD, 52, 71, 101-112, 131
  of PCDF, 51, 56, 71, 101-112, 131

Regulations, 101-112, 123-125,
  140-142, 226

Replacement fluids
  decomposition, 189-198
  selection, 147-150, 207-228
  toxicity, 225-226

231